Philosophical Foundations of Evolutionary Psychology

Philosophical Foundations of Evolutionary Psychology

August John Hoffman

LEXINGTON BOOKS
Lanham • Boulder • New York • London

Published by Lexington Books
An imprint of The Rowman & Littlefield Publishing Group, Inc.
4501 Forbes Boulevard, Suite 200, Lanham, Maryland 20706
www.rowman.com

Unit A, Whitacre Mews, 26-34 Stannary Street, London SE11 4AB

British Library Cataloguing in Publication Information Available

Library of Congress Control Number: 2015952408
ISBN: 978-1-4985-1815-4 (cloth)
ISBN: 978-1-4985-1817-8 (pbk.)
ISBN: 978-1-4985-1816-1 (electronic)

To Nancy, Sara and AJ: The Love of My Family
Has Defined the Meaning of My Life.

Contents

List of Figures

Foreword

Historically, the discipline of psychology has sought to investigate scientifically the issues addressing the origins of human behavior. Indeed, Psychology has, as its historical foundation, the basic concept of submitting classic ideas of philosophy to scientific scrutiny. For example, philosophers have long speculated (and debated) about the influences of genetics (nature) versus the environment (nurture) on all aspects of human personality and development. In Philosophical Foundations of Evolutionary Psychology, Hoffman attempts to draw parallels with these very ancient Greek philosophical views regarding the nature of human behavior with some of the more recent evolutionary interpretations addressing key terms such as evolutionary adaptation, natural selection and reproductive fitness. The etiology of controversial evolutionary topics, including the capacity of free will and human nature, altruism, and cooperative behaviors are explained through the beginning philosophical interpretations of Greek philosophers including Socrates and Plato. Cooperative behaviors, interdependency and human nature are interpreted as evolved mechanisms that provided the necessary framework in community development and social interaction.

In *Philosophical Foundations of Evolutionary Psychology*, we see an interesting and not previously presented comparison between ancient Greek philosophies addressing human nature with recent evolutionary topics addressing human behavior. The end result is a highly provocative interpretation of combining traditional theories of human nature with evolutionary psychology. Additionally, community development and connectedness are described as evolved behaviors that help provide opportunities for individuals to engage and support each other in a conflicted society. In sum, this manuscript helps piece together centuries old theories describing human nature with current views addressing natural selection and adaptive behaviors that helped shape the good that we know in each person as well as the potential destruction that we seem to tragically be witnessing with increasing frequency. At the time of this manuscript publication yet another mass campus shooting had occurred at Umpqua Community College (near Roseburg, Oregon). Human nature has evolved for literally hundreds of thousands of years with an inborn biologically-oriented need to cooperate with others in an environment that utilizes individual resources. Biology has helped shape who we are today as a species, but it has not become our ultimate destiny. Perhaps a better understanding of not just human nature but how human nature has genetically evolved through social interaction will provide us with the tools to establish a more cooperative and prosocial society that will reduce future antisocial behaviors.

Historically the discipline of psychology has attempted to uncover the questions addressing the origins of human behavior. For example, what are some of the contributing factors that have influenced both positive and prosocial behaviors that placed individuals at great risk and conversely what are some of the contributing factors that may explain human depravity and antisocial behaviors. In *Philosophical Foundations of Evolutionary Psychology*, Hoffman attempts to draw parallels with ancient Greek philosophical views regarding the nature of human behavior with some of the more recent evolutionary interpretations addressing key terms such as evolutionary adaptation, natural selection and reproductive fitness. The etiology of controversial evolutionary topics, including the capacity of free will and human nature, altruism, and cooperative behaviors are explained through the beginning philosophical interpretations of Greek philosophers including Socrates and Plato. Cooperative behaviors, interdependency and human nature are interpreted as evolved mechanisms that provided the necessary framework in community development and social interaction. In *Philosophical Foundations of Evolutionary Psychology*, we see an interesting comparison between ancient Greek philosophies addressing human nature with recent evolutionary topics addressing human behavior. The end result is a highly provocative interpretation of combining traditional theories of human nature with evolutionary psychology. Additionally, community development and connectedness are described as evolved behaviors that help provide opportunities for individuals to engage and support each other in a conflicted society.

Gary Starr
Psychology Dept. Chair
Metropolitan State University
October 2015

Introduction

Has Biology Become Our Destiny? A Critical Review of Philosophy, Evolutionary Psychology and Cooperative Behaviors

The field of evolutionary psychology to date has become one of the most important and timely descriptions and explanations of human behaviors and how humans have adapted and evolved over centuries. It is a science that has culminated over several centuries and currently presents itself as an inclusive and scientific exploration addressing the roots of human behaviors and explains human psychological functioning (Buss and Reeve 2003). Evolutionary psychology is unique in that it has been described as a science capable of drawing together other disciplines (i.e., psychology, anthropology and biology) in an effort to improve our understanding of human behavior (Liddle, Shackelford, and Shackelford-Weeks 2012).

A clear understanding of the current and historical development of evolutionary psychology through the centuries will allow us to understand how both the neurological processing of the brain and adaptive behaviors have evolved in response to specific threats to our well-being and related problems of survival. This text is most relevant today because it offers contemporary information pertaining to the evolution of human nature that can facilitate prosocial and cooperative behaviors as well as reduce potentially antisocial and destructive behaviors pertaining to both individual and group behaviors. Similarly, the purpose of this text is to provide graduate students and educators with a direct and scientific understanding of human nature in relation to the early history of Greek philosophy, modern (i.e., seventeenth century) psychology and more recently evolutionary psychology.

As David Buss (2005) describes it, "a robust field of evolutionary psychology has finally emerged . . . and Darwin's prophetic vision is now being realized—a psychology based on a new foundation" (p. xxv). In order to adequately understand evolutionary psychology as it presently exists, we need to explore how this discipline has evolved over the centuries from more basic philosophical theory that described human behavior. Several characteristics and domains of human behaviors (both positive and negative) have shaped the development of evolutionary psychology. For example, cooperation and aggression are two common behaviors that have had profound influences in how humans have interacted and survived (or perished) throughout history.

One of the most controversial topics throughout human history addresses the philosophical nature of the evolution of human behaviors—people have been described historically as either being genetically predisposed to violent and aggressive actions (Ainsworth and Maner 2012) or productive, compassionate and prosocial creatures (Goetz, Keltner, and Simon-Thomas 2010). Tragically, so often today we see (through the lens of social media) how destructive behaviors committed in the name of political ideology and religion (both domestic and foreign) can challenge traditional norms and perceptions of morality and threaten the very existence of civilized societies and communities. An analysis of how specific affective states (i.e., compassion and aggression) and social behaviors (i.e., cooperation) have evolved over time may provide us with important information in developing environments that may help prevent future conflicts.

Recent empirical sources today in community development, sustainability and evolutionary psychology argue that "connectedness" and empowerment are key components to a healthier (Wolff 2014) and less violent society (Cattaneo and Goodman 2015). A better understanding of how specific environments may (i.e., school and academic environments) contribute to people participating in a variety of community development projects that promote a sense of connectedness and belonging with each other may help to reduce some of these conflicts (Flannery, Liau, Powell, Vesterdal, Vazsonyi, Guo, Atha, and Embry 2003).

Before modern Humanistic psychologists such as Carl Rogers or Abraham Maslow discussed the positive characteristics of human behavior or Freud described the antagonistic qualities of the topographic and psychosexual theory of development, the earliest classical philosophers (i.e., Plato, Socrates and Descartes) were also deliberating over the essential qualities of human nature. They too were describing how behaviors became influenced through "fluids" or elements within nature or if behaviors where controlled by other metaphysical forces. The concepts of debate surrounding human nature thousands of years ago are strikingly similar to those theories still debated and discussed modern psychology today. Important topics to be discussed in this text include the controversy over "free will" versus innate of "fixed" behaviors, rational thought and the classic nature versus nurture argument, the true nature of the human personality as either collaborative and prosocial or egoistic and aggressive, the mind-body duality argument, and several others.

Understanding the historical influences of philosophy and evolutionary psychology can help to identify the causal factors of both constructive and prosocial behaviors that are necessary within our environment as well as the destructive and antisocial behaviors that tragically are becoming more evident in daily media. An important goal of this monograph is in providing the reader with an understanding how the classic philosophical theories helped develop the foundation from which modern psychology and most recently evolutionary psychology have progressed. While much of this text explores the relationships between classic philosophical and psychological theories, key human behaviors such as cooperation, altruism and aggression have been addressed as important factors contributing to human development and social interaction.

The concept of cooperation (a term that is used frequently in this text) is defined as the collaborative process of working with others for some type of mutually beneficial outcome; altruism is defined as some form of behavior designed to help others with no expectation of reciprocity; and aggression is defined as an intentional act to harm others. A related concept such as compassion is described as: "The feeling that arises in witnessing another's suffering that motivates a subsequent desire to help" (Goetz, Dacher, and Simon-Thomas 2010, p. 351). The topic of community and volunteer work can be applied to several different types of environments, such as working with community members in refurbishing a recreational park for children to play or simply picking up litter in a public area so all residents may live in and enjoy a cleaner environment (Hoffman, Wallach, Espinoza Parker, and Sanchez 2009). Empowering communities by providing opportunities for individuals to work cooperatively in a variety of service-related projects can help individuals not only feel a stronger sense of connectedness and belonging to those communities but can also improve cultural awareness and understanding (Jason 2006).

An important scope of this monograph is in exploring how concepts relative to human nature (i.e., cooperation and aggression) have evolved from the ideas of the early Greek philosophers and have influenced current views addressing psychological theories of evolution and human interaction. How, for example, could altruism and cooperative behaviors have evolved if early humans were considered primarily egoistic and opportunistic creatures, similar to the "survival of the fittest" culture we so often hear about? Clearly, humans must have evolved with at least the *potential* to engage in some form of prosocial and cooperative behaviors if our very survival depended upon it.

As Henri Parens (2014) describes human nature in *War is Not Inevitable*, war (and presumably conflict in general) is indeed avoidable depending on how cultures rear and train their children. Parens rejects the basic Freudian premise regarding the universal destructive human instinct (i.e., *Thanatos*) as well as the controversial notion of "moral relativism." Furthermore, Parens argues that individuals engaged in the practice of genocide and other horrific war crimes are by no means "normal" (p. 11) but rather psychopathological and that the ends does not justify the means no matter how noble our intentions are in achieving moral virtue. Parens (himself a World War II Auschwitz concentration camp survivor) encourages parents to teach their children to experience compassion and to exercise their free will in their social relationships with other children. He also encourages parents to teach their children empathy by reminding children what it must feel like to be the recipient of pain and torture. Parens would clearly agree, then, that in order for social justice and self-determination to exist within our society we as parents and educators must teach (by way of example) these values to younger populations. Cultural and community development in which individuals work cohesively together in achieving mutually important goals can not only prevent conflict (and war) but also improve understanding among diverse groups who comprise those communities.

NATIONAL TRAGEDIES: WHY DO THEY OCCUR?

On Friday, December 14, 2012, Adam Lanza had committed one of the worst mass murders in U. S. history at Sandy Hook elementary school in Newtown, Connecticut. Many theories and profiles have been proffered as a means of understanding how such a tragic event could ever happen, such as failure to recognize or understand feelings of others (i.e., "Theory of Mind") or an inability to differentiate feelings from the self with others ("mentalization"). Similarly, on Tuesday, April 20, 1999, Eric Harris and Dylan Klebold planted 99 improvised explosive devices in their Columbine High School and murdered 12 students and one teacher in Jefferson County, Colorado. Sadly, the shooting remains the deadliest mass murder committed in any high school campus on U. S. soil. More recently, on Wednesday, June 17, 2015, Dylann Roof shot ten individuals (nine victims died) at the Emanuel African Methodist Episcopal Church in Charleston, South Carolina. Roof was quoted as saying: "I have no choice . . . I am not in the position to, alone, go into the ghetto and fight. I chose Charleston because it is most historic city in my state, and at one time had the highest ratio of blacks to Whites in the country. We have no skinheads, no real KKK, no one doing anything but talking on the Internet. Well someone has to have the bravery to take it to the real world, and I guess that has to be me" (*New York Times* article, June 20, 2015).

Whatever was happening on these tragic days, what we do know is that Adam Lanza, Dylan Klebold, Eric Harris and now Dylann Roof all engaged in behaviors that had eerie similarities. Their behaviors matched other antisocial behaviors in the sense that their profiles reflect an inability to foster and develop communicative and connected relationships with their peers and community members. They were all loners who spent long hours on websites devoted to hate crimes, violence and racial segregation. Many of them spent significant amounts of time preparing manifestos *ex post facto* trying to justify their actions. Sadly, each had confided their horrific plans to a few acquaintances but surprisingly no one apparently thought that they would actually be carried out. There also appears to be consistency in alienation from larger social groups as a contributing factor to these tragic behaviors (Kim, Kamphaus, Orpinas, and Kelder 2011). Important questions still need to be answered, perhaps most importantly what were the indicators or "red flags" that we need to understand to prevent these actions from reoccurring? What can communities provide for individuals and families that may help to identify these kinds of tragic and destructive behaviors before they actually transpire? An important premise in this monograph is the basic and inherent human need to engage and collaborate with others within communities. Humans are indeed social creatures that have an inborn drive to share their skills and talents with others in creating a stronger, more resilient community. When deprived of opportunities to share experiences in identifying our similar needs and strengths with other diverse groups of individuals, ethnic polarization develops and ultimately human conflict.

In this text the classic philosophical theories as proposed by Plato, Socrates, Descartes, Locke, Rousseau (and others) are identified and described as having an important influence in the development of modern psychological theories describing human behavior and interaction. For example, Socrates argued that the concepts of virtuous and evil behaviors are not innate, and that all behaviors among humans are influenced by social and cultural values. Socrates argued that the greatest crime against human nature starts with ignorance and the uneducated mind. The important and relevant questions that Socrates posed over 2000 years ago are still quite relevant today: How can society and communities help deter antisocial and violent behaviors and promote social justice, compassion and equality?

Evolutionary psychology explores the characteristics of human development and adaptation with environmental and social factors in shaping and defining human behaviors. As science further investigates the causal factors of current human behaviors, we can also see how our evolutionary background and history continues to influence complex behaviors in modern society. We can also see how our past history and adaptive behaviors continue to shape behaviors in modern society. The information contained in this monograph is not sensitive to politically correct theories nor does it serve as an apologist for the more divisive and contentious topics addressing evolutionary psychology.

What it does propose is a bold and scientific explanation of human adaptive behaviors as they influence cognitive and psychological theory.

Evolutionary psychology has interesting parallels with famous theorists (both classical and recent) who have been no stranger to controversy: Socrates' suggestion that human behaviors may develop outside the influence of the Gods (a belief that argues humans can and should be responsible for their own actions and behaviors—the beginning argument of "free will" versus biologically determined behaviors); Sigmund Freud's reductionist views regarding human behavior; Konrad Lorenz and Niko Tinbergen's theory of ethology and adaptive behaviors among several species; John Watson's views of behaviorism and learning under a variety of different types of environments (i.e., Fixed Action Patterns). Historically the description of psychology as a science has fallen under the domain of interpreting behaviors as an interaction or function of environmental experiences combined with cognitive or thinking processes. More recently, however, the discipline and study of psychology has been described as a science that has had an early evolutionary background, dating back at least to 1859 with the publication of Charles Darwin's now famous *On the Origin of Species*.

More recent theorists have added considerably to the depth and understanding of evolutionary psychology. E. O. Wilson has developed an extensive analysis of social animal behaviors described in sociobiology, and William Hamilton's provocative views addressing evolution and inclusive fitness (i.e., the likelihood that offspring and kin will survive) have provided an extensive background to aid in our understanding of adaptive evolutionary human behaviors. Evolutionary psychology refers to the science of Darwinian principles in the process of our understanding human nature and human behaviors. Within evolutionary psychology, scientists explore the adaptation of key behaviors that enhance reproductive and inclusive fitness (i.e., the increased capacity to expand members of one's own species), *superfecundity* (the biological tendency to over-reproduce offspring), and our ability to pass on favorable traits that promote survival. In this sense, then, nature "selects" those traits and characteristics that proved essential to our own survival. Those surviving group members passed on key traits (i.e., intellectual traits) to future populations, thereby improving reproductive fitness within specific groups and contributing to their overall survival.

Perhaps the single most compelling and interesting aspect of evolutionary psychology is summed in one word: *Variability*. Variability can exist in several forms; both genetic and sociopsychological variations exist with important ramifications of community development. Genetic variation provided advantages for phenotypical adaptation to diverse environments, and sociopsychological variation provided individuals with different psychological traits to contribute socially to different communities. Without the development of genetic variation, our species would have failed to survive the harsh environmental elements thousands of years earlier. Darwin noted that the single most important characteristic in the existence of any species is not necessarily over reproduction (i.e., superfecundity), but rather the ability to adapt to changing topographic environments, climates, etc. Genetic variation has allowed humans and other species to adapt to the demands of the global environment through phenotypical traits. Darker pigmentation proved essential in warmer climates, lighter pigmentation proved essential in cooler climates, and those with reproductive and inclusive fitness were most likely capable of passing their adaptive traits on to future populations. These phenotypical traits defined by genetic variation were singularly the most important characteristic in defining human survival within harsh topographies and climates.

Most importantly, Darwin established that any given species is most likely to survive in a rapidly changing environment if the species retains the ability to adapt to changing demands within a specific environment (Darwin 1871, *Descent of Man*). Current research has identified three pivotal areas of Darwin's work that have had profound influences in the development of modern evolutionary psychology: Evolutionary change through natural selection, phenotypical change can be explained entirely through natural events within the environment, and extensive empirical evidence that supports these evolutionary changes (Dewsberry 2009; Buss 2009).

Additionally, the assimilation of both adaptive phenotypical traits (i.e., observable characteristics that define our physical appearance such as hair, skin and eye color) with psychological and social characteristics (i.e., communication, memory, collaborative and altruistic behaviors) proved essential to an organism's survival (human or otherwise). For example, Darwin noted in his original exploration of the Galapagos Islands in 1835 the distinct character and shape of the beaks of the finch bird which were unique to that area (Weiner 1994). These newer and odd-shaped beaks were shorter, blunter and more resilient to cracking than the other beaks of traditional finches. Darwin concluded that these distinct phenotypical beak changes in one particular geographical area were more than a simple coincidence. The characteristics of the physical environment (nuts with exceptionally hard surfaces due to extended periods of drought on the island) had gradually influenced which types of beaks were most adaptive and suitable for extracting foods that ultimately enhanced reproductive fitness among the finches. Over extended periods of time, nature had selected those traits that were most adaptive for survival (i.e., reproductive fitness) for the species of the finch birds. The process of natural selection had now been discovered.

Currently, many different types of behaviors (social, intellectual, cognitive and developmental) are being described as having an evolutionary and adaptive background and history. Aggressive behaviors, for example, may have clearly served an adaptive function (particularly among males) to protect offspring, secure fertile females for reproductive fitness (Ainsworth and Maner 2012), food or territory (Webster 2008). Similarly, facial expressions may have served as efficient universal indicators of emotional reactions to strangers who may have threatened our well-being (i.e., snarl or grimace) or those who we trusted (i.e., smile) (Ekman and Friesen 1986). Those individuals who were better equipped in accurately identifying the distinctions between welcoming smiles of strangers as opposed to threatening glares may well have been better suited to understand and recognize underlying motives of others, survive and pass their genes on to future generations. Today evolutionary psychology has roots in most fundamental aspects of human behaviors, such as physical attraction, mate retention and mating strategies, aggression, conformity, intelligence, language development, cooperative behaviors, ethnocentrism, and altruism (Badcock 2012).

How the early Greek philosophers viewed human nature and how behaviors may have become influenced through relationships with others within the context of groups and larger communities has important ramifications when attempting to understand complex theories involving the topics of modern psychology and evolutionary psychology.

This monograph will provide individuals with a clear and contemporary understanding of how the history of psychology developed from the early schools in philosophy dating back to Socrates and Plato and influenced the discipline that is now known as evolutionary psychology. It also provides the reader with a contemporary and alternative perspective to the development of key behaviors (cooperative and prosocial behaviors, aggression and antisocial behaviors) among groups that proved essential to human survival during human evolution. Understanding how human nature has been interpreted and defined throughout the last several centuries is an important and essential task that will allow us to better understand the origins of human behavior. Finally, this text will provide the reader with an understanding of the relationship between current themes of traditional psychology (i.e., "mind-body" duality, universal experiences of human emotions, etc.) and how these themes are linked and adapted to the development of modern evolutionary theory today.

Part I

EARLY PHILOSOPHICAL THEORIES CONTRIBUTING TO THE DEVELOPMENT OF MODERN PSYCHOLOGY

Chapter 1

Early Philosophical Theories in Psychology

PHILOSOPHICAL ORIGINS OF EVOLUTIONARY PSYCHOLOGY

Our world today is constantly changing and very complex. Similarly, human behaviors that have evolved remain complex in that we are capable of engaging in a broad range of behaviors that are either compassionate and prosocial in nature or deliberately destructive and antagonistic (Goetz, Keltner, and Simon-Thomas 2010). What are the social, cultural and environmental factors that interact to elicit such a broad range of behavior?

The purpose of this monograph is to provide the reader with an in-depth understanding of the relationship between early philosophical theories addressing human nature and how these theories have influenced the development of both modern psychology and more recently evolutionary psychology. Evolutionary psychology is a science that explores the etiology of current human behaviors and traits that have evolved (i.e., cooperation, altruism and aggression) as adaptive mechanisms within complex social networks that facilitated human survival.

Additionally, this monograph will explore important psychological factors that have evolved and have played an important role in the formation of cooperative relationships throughout human history. Finally, an important scope of the monograph is in examining how society and communities have evolved with the psychosocial mechanisms that can encourage positive (prosocial) growth and human development and help dissuade negative (antisocial) and destructive human behaviors.

The topics of psychology, philosophy and evolution are rapidly becoming exciting new areas of research given the fundamental and historical relationship that they share with each other (Buss and Reeve 2003). An accurate and complete understanding of modern psychology and how evolutionary psychology has most recently developed can only be accomplished with an understanding of basic philosophical theories addressing human nature.

How the early philosophers understood and characterized human nature as a distinct and rational quality had a profound influence in the subsequent development of psychological theory describing how human relationships could be both cooperative and supportive or antisocial and destructive. Additionally, an important scope of this text is to identify the need for diverse communities to provide opportunities for individuals to work collaboratively with each other that fulfills a very basic psychological and evolutionary need: Interdependency (Carvallo and Pelham 2006).

This monograph has been written in three parts: Part I (Chapters 1–4); Part II (Chapters 5–8) and Part III (Chapter 9 and Epilogue). Part I explores the early philosophical theories in psychology and how these theories influenced the development of evolutionary psychology. Part II describes the relationship between basic principles of natural selection and evolutionary psychology with current issues involving gender, aggression and cooperative behaviors. Finally, Part III of the manuscript explores the evolutionary history of cooperative behaviors and explains how this trait remains a necessary function within human society.

PHILOSOPHICAL ROOTS OF MODERN PSYCHOLOGICAL THEORY

Describing the qualities of human nature and behaviors have long been the topics of compelling interest among the early philosophers and scientists. More specifically, the topics of cooperation, conflict, greed, generosity and prosocial behaviors have shared a history of controversy in both science and philosophy (Gray, Ward, and Norton 2014). The controversy first begins through an attempt in understanding the basic elements of human nature, and whether our mind or soul is distinct from our physical body. The essential question has been (and remains) what is the essence of human nature? Plato argued that the mind is connected to and literally becomes a part of the soul, and for humans a lifelong struggle for rational, jurisprudent and virtuous behaviors was important for individuals to achieve understanding of the natural world. Just as characteristics of the natural universe operate through precision and natural laws, human behaviors become controlled through the logical operations of the soul. Plato was one of the first Greek philosophers to establish the importance of ethics and virtuous behavior as a key driving force that influences our relationship with others as well as a defining characteristic of inner harmony, wisdom and peace.

Plato argued that an introspective mind that questions and analyzes our own behaviors not only leads to personal insight and balance within our own lives, but this process of self-inquiry also provides us with compassion and "gentleness" in our relationships with others (Walsh, Teo, and Baydala 2014, p. 61). The discipline of philosophy then is quite important to communities in that they provide opportunities for individual growth and understanding, where values such as reason, ethics and respect are taught to each individual member.

In a recent article addressing the relevance of philosophical thought and logic in modern professions such as engineering ("*Fools for Tools: Why Engineers Need to Become Philosophers*"), John Kaag and Sujata Bhatia (2014) argue that many of the most fundamental principles taught by Plato and Socrates are still both necessary and quite relevant today. Soldiers, for example, need to be both skilled in the art of weaponry and defense but also first must be sound in their judgment of ethics, empathy and morality. Civil engineers, medical doctors, teachers and lawyers must first understand the distinction between efficacy of practice (i.e., "Do no harm to your patient") and prudent, ethical behavior.

Infrastructure that involves the use of bridges, space ships and automobiles that are designed with faulty materials as a way of saving money may have disastrous consequences for the public. Without understanding the important relationship between the virtues and ethics of human nature and human behavior, society ultimately becomes more vulnerable to abuse and exploitation—problems that unfortunately today are pervasive in the corporate world. Mark Freeman (2012) refers to this phenomenological capacity as recognizing (and simultaneously respecting) the social and intrapersonal need to understand the value and importance of engaging and collaborating with individuals (i.e., "Otherness") who otherwise would remain distant from us.

Plato's description of the "soul" was not considered so much of a spiritual component of the body (although it was considered to continue to exist after the mortality of our physical bodies) but more of an intelligent design that establishes mental and psychological states relative to our well-being. Thoughts that govern our emotions and behaviors (i.e., happiness, sadness, anger, etc.) are first directed through the mind which in turn becomes influenced by the soul. However important rational and prudent thinking was according to Plato and other Greek philosophers, humans are still often mislead or "trapped" by their senses that often do not accurately portray the true characteristics of the physical world. Our physical senses (i.e., vision, hearing, touch and smell) can often misperceive actual elements within the physical world that prevent us from understanding reality as it truly exists. Similarly, our physical behaviors may have become influenced through "implicit associations" (unconscious states that indirectly exert influence over our actions) which we purportedly have very little control over (Stark 2014). These limitations of our own perceptions (and our ability to measure our own moral assessments) betray our ability to understand the elements of the natural world and understand the laws of the universe. The intense, bright and dazzling sun that exists in the world from Plato's *The Republic* stuns us and subsequently prevents humans from directly comprehending the world as it truly exists.

In the famous allegory of the cave in Plato's *The Republic* (380 BC), the ability to accurately perceive "reality" or the outside world of the cave becomes transfigured through shadows that represent real images outside the cave. The overpowering bright light from the outside sun prevented humans from not only seeing objects within the physical world as they exist but limits our understanding of how things operate within our world. Plato argues that humans have an inherent inability to accurately perceive reality as it truly exists—we are limited to interpreting behaviors from mere shadows in the outside world.

For humans, the ability to understand the true qualities and characteristics of the physical world continues to be challenged by subjective interpretations through our senses and cognitive representations (Machery 2011). Plato describes early childhood as a type of "binding" or limitation to an accurate understanding of the characteristics of the natural world because of the lack of experiences in the world and ego-centrism. Knowledge and a true understanding of the world are only achieved through a process of self-inquiry and critical thinking. But in order to understand the psychology of human behavior, a clear and accurate understanding of the relationship between science and human nature itself is required. The philosophical interpretation of one's life may start with the meaning or purpose of life, but the *biological* purpose of life itself is reproduction and sex. Without vast reproductive capacities, life as we know would cease to exist.

Much has changed since the early Greeks interpretation of knowledge and how it influences human behavior. But one thing remains the same since Plato's and Socrates' early descriptions of human nature and knowledge, and that is the scientific responsibility to understand the relationship between cause and effect, the scientific method and the value of introspection. Socrates also abhorred both arrogance and ignorance, citing the fact the evil behaviors stem not from innate causes but simply the lack of an examined and challenged (i.e., educated) mind. Science and empirical data begins simply with observations of events that transpire within our natural world. However, because our perceptions of observations and physical events can often be very misleading (Loftus 2013) and indeed subjective (i.e., "shadows" may hide reality), a process better designed to understand qualities of the physical world gradually emerged that is now referred to as the "scientific method." It wasn't until 1859 when *On the Origin of Species* was published that observations of the physical characteristics of plants and animals were introduced that gave us a more accurate account of the origins of humans as a species.

Darwin noted in his detailed observations from a variety of species on the Galapagos Islands that those animals with a greater ability to adapt to the changing physical demands of the environment through genetic variation had a greater chance of survival, and those traits of the surviving group were passed on to future generations (a term that Darwin refers to as reproductive fitness). Species that had the ability to reproduce in vast numbers via genetic variability within a highly diverse environment and adapt to changing environmental conditions had a significant advantage against other species relative to their own survival. This was not an issue of "quality over quantity" but rather "quantity over quality" where the more one individual can reproduce the better his or her chances are in surviving against a multitude of often fatal and predatory elements. Similarly, Darwin (1859) noted early in his travels off the coast of the Galapagos Islands that the majority of mammals tended to over-reproduce: "Generally . . . the most vigorous of males, those which are best fitted for their places in nature, will leave the most progeny" (*On the Origin of Species*, p. 136).

This advantage provided individuals from different species the capacity to survive by meeting unique demands imposed upon them from unique or foreign environments. The purpose of evolutionary psychology is to provide a rational explanation of a diverse range of human behaviors and psychological mechanisms through adaptive behaviors that have evolved (and continue to evolve) over thousands of years.

Additionally, those individuals better adept at mate retention showed greater ability in passing their genes to future generations (i.e., reproductive fitness) than those not possessing this trait (Buss and Shackelford 1997). Yet, given this fundamentally profound and biologically oriented human trait, it should not be surprising that historically so many high profile (and less prolific) individuals have risked their careers, families, and reputations when accused of unethical or "sexually inappropriate" behaviors (i.e., Most recently 2012 Four Star General David Patraeus and biographer Paula Broadwell; 2004 Vice President Candidate John Edwards and videographer Rielle Hunter).

These infamous behavioral characteristics have existed for centuries and have traditionally superseded political or religious identification and economic status. According to current theories addressing reproductive fitness and *superfecundity*, these behaviors can at least be understood through evolutionary psychology and "inclusive fitness." Ironically, one of the most profound and dynamic of all topics addressing evolution, sexuality and reproductive sciences has traditionally remained one of the most taboo and controversial topics throughout literature reviews. These topics remained taboo for most of the nineteenth and twentieth centuries due to threats of political pressure and censorship to scientists who were practicing in those fields.

Humans have always sought a deeper understanding into meaningful and purposeful events as long as records have been kept documenting human behaviors. Historically the aim of most traditional psychology texts has been to accurately describe the meaning and purpose of human behaviors through cognitive, social and intellectual development based on environmental experiences. More recently, evolutionary psychology has emerged at the forefront of science in identifying both genetic and evolutionary factors as key elements in understanding human behaviors. Several important contributions have emerged as a result of evolutionary theory in psychology:

- Evolutionary psychology is considered to be a uniquely important contribution to psychology based on the works of Charles Darwin, Alfred Wallace, William James, Herbert Spencer and more recently Richard Dawkins and David Buss;
- Evolutionary psychology provides the frame and fundamental structure that now integrates all major theories in psychology (i.e., developmental, biological, etc.) that attempt to explain human behavior;
- Evolutionary psychology provides an understanding of how persistent modern behavioral problems (i.e., aggression, overpopulation, *superfecundity*, polyandrous/polygamous behaviors, greed and antisocial behaviors) may logically be explained via adaptive behaviors from our past history;
- Evolutionary psychology has historical significance, as it has integrated several other sub-disciplines within the broader field of psychology to create a coherent tool in improving our ability to understand human behaviors (i.e., Functionalism, Developmental Psychology);
- Evolutionary Psychology provides rational (and often controversial) explanations of a broad spectrum of human behavior. It does not (as some critics have claimed; see for example Kruger, Fisher, and Wright 2013) attempt to "justify" unethical or criminal behaviors (i.e., sex by force as a means of improving reproductive fitness and lineage) but rather provide empirically grounded science in the hopes of an improved understanding of human behavior.

Evolutionary psychology has had a direct impact on changing how scientists have viewed the development of key human interactive behaviors, such as communication, cooperation, altruism, aggression, attraction and interethnic cooperative behaviors. Evolutionary psychology has emphasized the role and functions of our behaviors in social groups as opposed to more traditional introspective roles of behavior (Confer, Easton, Fleischman, Goetz, Lewis, Perilloux, and Buss 2010).

Since at least the seventeenth century (and even earlier), psychologists have attempted to describe behaviors as they exist within our environment as having some form of an adaptive and meaningful value that facilitated human existence (Wundt 1904; James 1950). Psychologists and philosophers have disagreed, however, about the impact and the distinctions between topics addressing the physical body (i.e., Materialism) and the possibilities of a higher order of human functioning (i.e., Idealism and Spirituality). Although earlier philosophers attempted to explain human behaviors purely from a scientific perspective without influence from religious dogma (i.e., August Comte's views of Positivism), Darwin's views of human behaviors were unique as they described the functions of human behaviors to be primarily a result of environmental influences and genetic variation. Differences in physical appearances as well cognition and intelligence were attributed to a variety of factors, both genetic and environmental that influenced how humans interacted with each other.

MIND-BODY DUALITY & WHAT IS THE MEANING OF OUR OWN EXISTENCE?

For the first three billion years of our own existence, humans really didn't need to differentiate between such aesthetic and cerebral topics as the distinction between "ideas" and "reality" or ponder the meaning of our own existence. To have the ability to do so is actually a luxury—luxury in the sense that humans could afford to spend their time not engaged in foraging for food or fending off predators. It was not until the fourth century (i.e., Plato) were philosophers capable of engaging in cognitive or abstract critical thinking and devoting their time to things other than life-threatening situations.

Perhaps one of the most controversial topics within philosophy and psychology was the question of the duality of the "spirit" or the mind and the physical body. To what degree are our thoughts separate and remain uninfluenced from physical properties? Can all elements within the natural world be explained through physical phenomena? Can humans objectively measure and quantify their own experiences within the environment? Are our thoughts and functions of the mind capable of being analyzed into separate units?

Or perhaps most importantly—what is the overall meaning and purpose of our own existence? Plato argued in some of his earliest works (i.e., Plato's Dialogues) that ideas existed within the soul prior to the development of the mortal self.

The mind existed separately from the body and continued its existence after our own death. This theoretical perspective argued that thoughts and ideas were separate from our physical development. Additionally, Plato argued that our "ideas" which are contained in the mind continue to exist to develop and grown even after our death.

An important outcome of the development of evolutionary psychology and natural selection was to actually help *close* the divide among psychologists in various sub-disciplines (i.e., biology, anthropology) by identifying a common (i.e., monogenetic) origin of humanity (Charles 2013). Although today the discipline of psychology may appear to be split and even disjointed based on differing perceptions of the causality of human behavior, this was not always the case. Questions regarding the nature of human consciousness (phenomenology) and knowledge (epistemology) were initially addressed by philosophers and earlier schools of psychology (i.e., Structuralism and Functionalism) and gradually became diverse as different theorists challenged basic interpretations of "consciousness" and "experience" (Charles 2013). An important scope of this monograph is to show the reader how early philosophical interpretations of behavior and consciousness have influenced theories of psychology and how modern psychology has actually supported the basic tenets of evolutionary theory, natural selection and adaptation.

THE ZEITGEIST OF PSYCHOLOGICAL SCIENCE

Ever since recorded history even began, humans have attempted to help other humans who were experiencing some form of physical or psychological impairment. The "help" that was provided created a close network of relationships with clans that made development and survival possible. However, throughout centuries and generations each group had unique methods and styles of addressing physical and psychological problems. The term "zeitgeist" refers to the cultural and social influences that shape values that are practiced within any society. The term zeitgeist, literally defined, means "spirit of the times" and is a critical component and topic in the historical development of the discipline of psychology that often determines how a particular academic or scientific discipline progresses in terms of research, scholarly activity and productivity. For example, a culture that remains conservative in its approach to scientific exploration and theoretical development may limit progress and innovation in developing new theories relative to their science. Those cultures embracing a more conservative zeitgeist may even continue practices that are actually dangerous or harmful to patients.

Conversely, the zeitgeist of other more progressive cultures may actually encourage development of theories relative to their discipline, thereby expanding and proposing new concepts to further knowledge and information that ultimately helps people. Scientists, researchers, and educators may be somewhat reluctant to propose new and controversial ideas (i.e., Darwin's views of evolution directly contradict the teachings of many religions) as the repercussions of such actions may have jeopardize their work, reputations and in some cases event their own safety. Even today some educational institutions of higher education may reject current theories addressing the etiology of humans and subsequently progressive theories in science explaining human behaviors because of perceived contradictions between traditional views of theology and modern views of science. The historical development of psychology has traditionally been influenced by several factors, including political climate, social and religious factors as well as the basic temperaments of populations within any community. The term "zeitgeist" is a concept that is often used to describe how a particular attitude or cultural belief system may influence not only how psychology is explored and investigated, but indeed how psychology as a science defines itself. Ignorance, prejudice and discriminatory belief systems may have delayed or prevented the scientific exploration of subjects that even today have become highly divisive and contentious.

VICTOR FRANKL AND HENRI PARENS: WHY CONFLICT (AND WAR) IS NOT INEVITABLE

Traditional research in psychology has described the human personality from a more inherently destructive and highly antisocial perspective (Freud 1930). Rules and laws were essentially created to prevent further human aggression and violence due to the inherent human biological and aggressive instincts. More recently, however, theorists have identified a more positive side to human nature, one that describes human potential more from the "strengths perspective" rather from the "deficit perspective." Many sub-disciplines within psychology (i.e., community psychology) have primarily emphasized individual human strengths, competence and aptitudes despite the challenges of social injustice and oppression (Elias 1994). Additionally, current psychological research supports the role of spirituality and positive emotions in improving not only our psychological framework in our relationships with others (Gruber, Quoidbach, Kogan, and Mauss 2013), but can also improve our physical health when recovering from potentially life-threatening illnesses, such as heart disease and breast cancer (Mouch and Sonnega 2012; Gall, Kristjansson, Charbonneau, and Florack 2009).

In his interesting analysis of human nature, conflict and aggression, Henri Parens (2014) addresses the question posed by Albert Einstein to Sigmund Freud in a letter dated July 30, 1932: "Is there any way of delivering Mankind from the menace of war?" Parens argues that despite human nature's tendency toward continuous conflict and aggression, war is clearly not inevitable if we can merely address some of the important psychodynamic factors that contribute to human behaviors. Parens challenges the traditional Freudian interpretation of aggression as something that is fixed and innate (i.e., Thanatos). Parens offers hope for the Human condition in that human aggression is learned through our relationships with others, and that certain styles of parenting (authoritarian parenting styles and "unquestioningly obedient children" [p. 236] as contributors to the brutal and inhumane behaviors demonstrated by Nazi concentration camp officials during World War II).

Parens also argues that it is possible for children and adolescents to become morally responsible and humane individuals through the context of authoritative parenting styles that provide secure attachment formations with caregivers. In a word, Parens argues that parents need to display those desirable behaviors as role models if we wish to end conflict and antisocial behaviors.

A second major literary figure who rejected Freud's theory regarding the inevitable nature of human aggression was Victor Frankl. Victor Frankl argued in his now famous text: *Man's Search for Meaning* (1959) that harsh and life-threatening conditions in a Nazi concentration camp (Auschwitz) still could not prevent him from living a life of meaning and purpose nor would he allow the German soldiers to rob him of his will to live. Frankl argues that happiness in life is essentially a byproduct and that all persons need a driving force or a reason to survive to achieve meaning and purpose in their lives (in Frankl's case, his will to live came from a hope in seeing his beloved family members once again—tragically this never transpired for him as they were executed in Dachau and German concentration camps shortly before the end of World War II).

Shortly after his release from the Nazi prison Frankl published a series of articles describing the relationship between meaning in one's life, experiential processes in our relationships with others, and problems associated with cultures and society that are preoccupied with materialistic possessions (i.e., conspicuous consumption) criteria for a successful and happy life. Cultures that value possession and ownership of items as criteria of success and those that emphasize physical characteristics of beauty as a means of achieving happiness are too often comprised of individuals suffering from emotional and psychological problems (i.e., depression). Additionally, Frankl was able to incorporate Nietzsche's often quoted phrase: "What does not kill us makes us stronger" into his own theories of Logotherapy to help people confront meaningless pain and suffering in their own lives. Having a purpose to survive within even the harshest and most dismal of environments will provide us with the insight that through pain and suffering we can find meaning within our lives.

Religion and spirituality were frequently described by the early philosophers as being an inherent component of human nature that enhanced our own existence and differentiated humans from other mammals. In order to understand the characteristics of the physical body, individuals were required to understand the metaphysical components (i.e., spirituality) of human life. Animals had no soul and therefore were incapable of experiencing divine transformation or metaphysical properties. According to the early Greek philosophers, humans are obligated to pursue and investigate the mysterious spirituality of human nature. Without an adequate understanding and awareness of our more spiritual and aesthetic historical background, our lives would be considered meaningless (i.e., imbalanced) and capable without reason and rational thought.

Both Frankl and Parens argue that while conflict is unfortunately a very common event among humans, it clearly is not inevitable. Communities that are designed to provide individuals with opportunities for growth, mutual understanding and shared resources are effective in not only reducing potential conflict, but they are also highly effective in helping individuals to discover similar characteristics that we all share together. Superordinate goals, community service projects and a variety of other activities that can actually bring individuals together will not only create positive relationships among individuals, but can provide strength in community development (Putnam 2000).

Chapter 2

Primary Philosophical Influences in Psychology

SOCRATES (469–399 BC)

An important and controversial topic among the early philosophers was that of human nature and how human nature shapes and guides human behaviors. How people socialize and interact with each other and to what degree are behaviors egoistic and self-serving as opposed to selfless and prosocial behaviors directed to the good of the community were central topics of debate. Several of the classic Greek philosophers argued that knowledge is innate and that our ability to understanding what is "right" from "wrong" essentially depends on self-reflection and introspection. Perhaps no other Greek Athenian philosopher has had more of an impact in literature, science and western cultures than Socrates. This is in fact ironic given the fact that Socrates never actually has been noted to record any of his works. Really most of what we know today about Socrates comes from his students and a few devoted followers: Plato (*Apology*), Xenophon (*Symposium*) and one parody of a play that portrays Socrates as a clown by Aristophanes (*The Cloud*).

The Socratic Method and the Scientific Method

If information comes directly from asking questions to ourselves, how are we to obtain any kind of new or insightful information? Socrates would answer this question by arguing that knowledge from other persons is useless, and that obtaining information by ourselves and for ourselves is tantamount to knowledge and virtue. Knowing what is "Good" and knowing "Justice" can only come from critical self-analysis and through a detailed process of asking ourselves questions, which ultimately should provide ourselves with the answers that we are seeking: "I know that you won't believe me . . . but the highest form of Human excellence is through the process of questioning oneself and others" (www.philipcoppins.com/socrates.html). Socrates questioned the idea that some truths (i.e., virtue) could even be taught. This criticism supports the concepts of a much later philosopher John Locke who compared our ability to learn to that of a Tabula Rosa or blank slate. Knowledge, therefore, is learned and prone to be individualistic based on the unique experiences that we share within our own environment. Several individuals experiencing the same event within their environment or a particular type of behavior may all focus on different events, thereby perceiving events differently but viewing the same stimulus simultaneously. Socrates notes that the common argument between the origins of "good" versus "evil" is misguided. No person, according to Socrates, is inherently "evil" so much as they are ignorant. Socrates (and later John Locke) argued that human nature is neither redeeming nor negative but merely a reflection of the values that were taught to the individual by society.

In modern terminology, communities and cultures within society help shape and govern individual behaviors by portraying role models in desirable ways that appeal to younger persons or in a more detrimental way that has negative consequences (Bandura 1986). We cannot blame individuals for engaging in corrupt or antisocial behaviors when they have not been properly taught of the consequences of such behavior. Thus, Socrates would explain even the most diabolical and destructive behaviors (i.e., Adam Lanza of the 2012 Sandy Hook Elementary School shootings or Dylann Roof of the 2015 Emanuel African Methodist Episcopal Church shootings) would be due to serious problems relative to specific cultures and learning environments that are typically associated with teaching and training children morally just and ethical behaviors. In this way, then, Socrates' views regarding individual behaviors are a function of the culture and environments that inspire children to learn and understand the relationship between their thoughts, behaviors and individual happiness.

Socrates most likely would agree with John Locke that ideas of "good" and "evil" or just versus unjust behaviors are in fact self-imposed and culturally influenced doctrines that individuals learn from experiences within their environment (Locke 1963). Communities and cultures that emphasize emotion-related cognitions (i.e., positive affect) have been shown to be significant predictors not only for mental health but also for physical health (Pressman, Gallagher, Lopez, and Campos 2014). Numerous empirical studies have shown the relationship between affect (positive or negative) as characteristics that become influenced through culture and ultimately influence well-being and health (Curhan, Sims, Markus, Kitayama, Karasawa, Kawakami, Love, Coe, Miyamoto, and Ryff 2014). Values are reflected from our culture and are taught to individuals within the community—it is precisely these kinds of cultural values that shape and determine individual behaviors. Thus, individuals who are exposed to a collectivistic culture may be more inclined to implement cooperative principle-sharing resources because these key values are emphasized within specific cultures.

Socrates was primarily arguing that knowledge, wisdom and insight are both individual and collective processes that are learned within the culture and environment in which he or she is exposed to. Our thoughts and behaviors are established first within our culture and the context of our relationships with others that can also influence our physical health. Learning therefore becomes an experiential process that influences both mental and physical health.

The majority of the philosophers previously discussed would argue that if *individual* behaviors need to be changed, then an examination of the society and culture from which these behaviors develop also require examination. For example, Socrates argued that while he could not tell others what is inherently good or bad, he would challenge them to question their own world views that so many people accept automatically (i.e., religious views, attitudes toward government, democracy, etc.). Virtues such as insight, knowledge and truth can only be obtained through living a life of self-questioning and contemplation. Truth, therefore, is neither "taught" nor learned but discovered within oneself through a lifelong process of self-inquiry.

It is no wonder that eventually Socrates was charged with treason and "corrupting the minds of youths" (i.e., questioning the divine authority supposedly held by the gods of that era) in Athens which eventually led to his indictment and death sentence of dying by hemlock. His basic crime was in teaching Athenian youths to "question authority" shortly after the Peloponnesian War (431–404 BC) and to question what the term "enemy" actually refers to, and asked if "God" was really on the side of Athens. Additionally, Socrates sealed his fate and charge of treason when he actually refused to assail the Spartans as "the enemy" and argued that perceived enemies are no different than our perceptions of ourselves. Socrates was better at understanding how different groups have more similarities than differences, and frequently challenged the justification of any war. He helped resolve conflicts and differences among groups of individuals by identifying the common ground that we all share as humans and focused on common areas of interest (i.e., truth and insight) that benefit communities and society. Perhaps the popularity of the philosophy of Socrates was that he argued that all persons are capable of achieving knowledge and insight simply through questioning themselves and assuming that they do in fact have things to learn about the world around them.

Socrates and the Scientific Method

What is now perceived as a basic foundation in experimental psychology and methodology in modern psychology is the scientific method. Socrates is often described as one of the founders of the scientific method as his method of questioning individuals was actually a method of identifying through the process of elimination hypotheses. More recently other philosophers such as Francis Bacon (1561–1626) and later John Stuart Mill (1806–1873) have been identified as the primary proponents of the scientific method and empirical thought in science. The "Scientific Method" is typically described as a process of extracting or obtaining information and follows four steps:

- Knowledge of a general idea or conditions—What is "Knowledge"? or What is the relationship between two or more variables? For example, what is the relationship between physical activity and moods?
- Formulating a testable statement that is based on a theory (i.e., a hypothesis); Question yourself and your behaviors—Look for contradictions in what you say and what you actually do. An example of a testable hypothesis might include a specific statement from the general ideas listed above: 30 minutes of physical exercise increases levels of happiness;
- Testing the idea under controlled conditions (i.e., a controlled environment, such as a laboratory) that is commonly referred to as an experiment; and
- Interpreting the results to determine if the original hypothesis has been supported or discounted. Socrates' experimental method was through a process referred to as "mirroring"—how can our behaviors match our personal beliefs and doctrines that exist in society and culture (see Figure 2.1 below).

Humans as Cooperative and Rational Creatures

Socrates argued that wisdom, virtue and truth are only obtained through critical introspective analysis of the self. Knowledge is gained through the process of questioning one's self. These qualities are *apriori* in the development of a noble and virtuous life. The question of what is "good" became rather irrelevant to Socrates as he was more concerned with *how* we propose to live a good and justify our virtuous lifestyle. These qualities cannot be taught by others but are learned through a meticulous process of self-inquiry and autonomous thought processes. In terms of community development, what is determined to be virtuous behavior can typically only be determined within the context of group and social interaction (i.e., personal sacrifices for others). Communities and groups are better served when individuals are encouraged to think through meaning and purpose of their own lives. It is precisely through this process of self-inquiry that we achieve meaning and purposes within our own lives and are thus better capable of becoming responsible and ethical citizens. If a community can provide methods of self-analysis and critical insight, then the community itself provided citizens with knowledge that was capable of being contributed to the greater good of society.

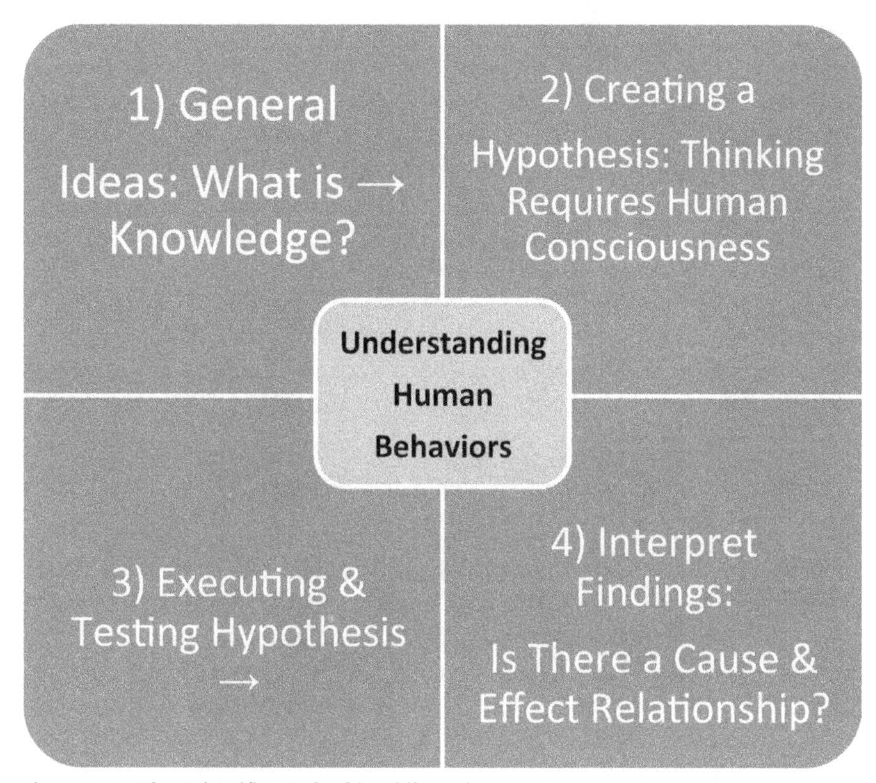

Figure 2.1 The Scientific Method in Philosophy.

PLATO (427–347 BC)

A second and arguably more famous Greek philosopher who has had a pronounced impact regarding the development of modern psychology was Plato. Plato is perhaps best known for his works addressing the topics of "Moral Psychology" and the need to place our own self-interests first above a priori others. Plato also argues that distinctions exist between our *perceived* ideas of self-interest and the greater needs of society in general. Ironically, what we now know mostly about Socrates comes directly from his students Plato and Xenophon (Socrates never actually wrote or even recorded any of his philosophical insights, and left this task to his student philosophers).

Like Socrates, Plato argued that all of our ideas are inborn (innate) and that there is no division or distinction between thoughts, the mind and our physical being. "Ideas" and "truths" exist in their purest form *prior* to the formation of the physical or mortal body. The mind continues to exist even after our physical bodies no longer live, and these ideas continue to develop.

In order for rational and coherent thought to exist we must consider our own needs and desires first. Our ability to refrain from (or at least delay) physical needs and pleasures provided the intellectual impetus to think more critically and see natural events within our world more accurately.

Perhaps one of the most famous proponents of the Idealistic philosophy (i.e., what humans are *capable* of achieving) was Plato. Plato's perception of the universe has been described in three basic forms: Our vision of the Ideal World (perceiving elements as they exist in the universe); the second perception includes the Spiritual World as defined and created by God. These elements defy human interpretation; and the third form is what Plato refers to as the Psychological. The Psychological form of our world refers to our mental interpretation of events occurring in the physical world.

Cooperative Behaviors as a Critical Component of a Productive Society

If we were to interpret these classifications of worlds today, the Ideal World as described by Plato would be one where people work productively and cooperatively with one another as a means of reducing common social and economic problems—one may even consider this a modern-day "Utopian Society." Plato argued that personal vices, such as greed and violence are a function of a mind that has not been fully examined (Goodey 1992). Rational and temperate behaviors were outcomes of a thorough analysis and understanding of one's mind (Simon 1972). Humans are capable of achieving a greater level of cognitive development through rational and intellectual thought. We would be able to understand and perceive nature is it exists in the world and subsequently addresses problems that are associated in the physical world. "Happiness" as a mood state can be developed and maintained through a variety of community activities and interaction with groups of individuals that share a common goal as we do. As a highly subjective concept, happiness can be experienced for duration despite numerous social, physical and economic challenges that individuals commonly experience (i.e., poverty and oppression, even imprisonment).

The Platonic views of happiness and spirituality provide explanations how individuals may experience elevated moods in their work with others and may even serve as a coping mechanism when exposed to chronic stressful environments (Pargament 1997). Happiness, then, is an abstract concept proposed by Plato that is achieved primarily as a consequence determined through our relationships with others. According to Plato, the Ideal World is one which we as humans are capable of achieving if we combine our intellect and cooperate with each other, and the second world is the Spiritual World. The Spiritual World questions whether we as mortals are capable of understanding Divine Inspiration and the topics of God. Historically, many philosophers even before Plato argued that the single most important dimension of human consciousness and behaviors is the ability to understand the existence of a spiritual being commonly interpreted as religion and theology.

Moral behaviors can exist (even altruistic behaviors) according to Plato, but individuals within a society must first recognize and attend to his or her needs before recognizing the needs of others in society. In perhaps his best known publication, *The Republic*, Plato argues that the ideal characteristics found in all societies are inherently relative to one another and are contained (i.e., as traits) in each person. Having traits does not mean demonstrating these traits, and both Plato and Socrates argue that the responsibility of the state is to create societies that allow these noble behaviors to manifest themselves throughout daily interactions with other members of society.

Socrates argued that the three great virtues: Temperance, Wisdom and Courage are sustained through the abstract quality of justice. Socrates, for example, also argues that "just" behaviors are essentially equivalent to those behaviors that conform to the laws of society or "higher authority." The best rulers for any city, state or government are not politicians but rather philosophers who should be reluctant to take on such a powerful position. Additionally, Socrates himself claimed never to "knowingly commit an injustice" (Johnson 1990), and that the defense of those in need and helping friends in great distress comprise the mechanisms of justice. Both Plato and Socrates would also argue that communities function most efficiently when people are of equal status (i.e., men and women having equal rights), all property belongs to public access (i.e., no private property), and that the interests of the group are best served when justice exists for all persons who comprise the group.

In this sense, then, even the earliest of Greek philosophers such as Plato and Socrates have argued that certain psychological characteristics are ideal in group situations and perhaps most importantly that all individuals are indeed capable of demonstrating these behaviors under the right environmental conditions. This philosophical perceptions has influenced psychologists today in the "nature—nurture" controversy in the sense that environmental conditions, styles of learning and rewards and punishments can elicit or emit appropriate behaviors within a variety of different types of contextual environments. Plato felt that the mind and physical body are in fact separable entities and that the mind continues to flourish and exist after our own mortal death has occurred.

These psychological characteristics that were first identified by the Greeks were important in the development of psychology as now humans were considered to be individuals with behaviors that may have had predisposing traits or characteristics. Characteristics that may have defined adaptive behaviors, such as intelligence, physical strength and language capacities were now viewed as critical elements that not only shaped our ability to survive in a different environments, but perhaps more importantly these traits were viewed as being transmitted to future populations to further help in the challenge for survival. The Greeks also argued that behaviors designed to strengthen societies and communities via personal sacrifices were ideal and that values need to be taught (i.e., humility) to future generations.

The beginning development of the benefits of cooperative behaviors as a group asset in helping societies and communities to strengthen each other actually developed from the basic tenet's of Plato's Ideal World, dating back to approximately 400 BC. Platonic Idealism is described as both an abstract and basic concept where humans have the capacity to alter and change conditions within their own society that improves the many important characteristics of living and psychological well-being. Plato argues that the mind is the cause and influence of all behavior and the soul provides a rational and cognitive basis from which decisions are made.

Although there have been several different types of psychological views regarding human behaviors, the central concept and purpose of psychology has primarily remained quite consistent: To provide individuals with a clear understanding of both human behaviors and the thought processes that are associated with those behaviors. Plato was among the first philosophers to argue that human behaviors typically vary due to the content that was contained within their souls. Both Plato and Socrates would agree that during mortal life humans have the social capacity to cooperate among one another and share resources to improve their overall quality of community living.

RENE DESCRATES (1596–1650)

An important influence in the development of modern psychological thinking was Rene Descartes. Descartes argued against some of the more fundamental concepts and dogma of the Catholic Church and argued with other current philosophes of his era (i.e., John Locke) that knowledge is essentially derived from our experiences within our environment. Descartes as a young man was not considered "ambitious" and spent most of his youth actually in poor physical health. Most of his work and writing was actually done while still in his bed. After several brief temporary romantic relationships Descartes decided to commit his life to the application of mathematical principles and human development. He wrote these views in his text: "*Spirit of Truth*." Descartes argued that human behavior was comprised essentially of two basic elements: The Mind (i.e., thoughts, cognition and reason) and Body (i.e., our physical behaviors). The primary emphasis in "*Spirit of Truth*" was that science and mathematics are topics that could be implemented to better understand the basic principles of human behavior. Descartes also emphasized that the mind is not a type of physical substance, but rather comprised of non-physical substance (the mind existed as part of the soul).

Descartes further identifies two different types of substances: Cognitive (i.e., thoughts) or mental and substantive or physical (i.e., matter). These views later became known as the Cartesian Dualism principle. Traditional views of duality of human behavior have argued that the mind primarily influenced the physical body and related environmental concepts, such as motion or movement, coordination and reproduction. Descartes' views challenged those traditional aspects of duality and proposed that first psychophysiological approach to human behavior: A reciprocal relationships exists where our thoughts and beliefs influence our physical state of well-being. If our thoughts are about anxiety and stress it is logical to assume that the physical state of our bodies will respond in a similar manner.

Because the physical properties of body are made up of matter, human behavior must be subject to laws of nature and physics and is therefore predictable in a mathematically measurable world. Descartes should be noted as one of the first philosophers to describe the reciprocal relationship between how our thoughts influence physical aspects of the human body and introduced the concept that is most commonly known today as duality. Descartes argued that while the brain primarily controls cognition and intelligence, this universal function should not be confused with the nonmaterial existence of the mind, self-awareness and consciousness. According to Descartes, to be "self-aware" implies that we are cognizant of our own capacity to think and that thoughts reflect mental activity. This capacity is distinctly human and separates us from plants and animals, both of which lack souls. Descartes believed that some of the higher functioning animals were capable of limited cognitive functioning, and that behaviors of animals were essentially reflexive in nature caused by what he referred to as "animal spirits." Sensation, such as light is a form of energy that stimulates specific portions of the brain (occipital lobes) that helps us to process meaningful information pertaining to our environment. However, Descartes also suggests that these impulses from the environment are then sent to the spiritual world that exists beyond our comprehension.

Prior to Descartes, most philosophers (i.e., Plato) had argued that the mind and body were separate entities and that the mind continued to exist even after our own physical death. Earlier philosophical beliefs dating back to Plato had argued that the mind had little influence over the body and that the mind continued to exist well into the Afterlife. Socrates had mentioned in earlier writings that his behavior was often guided by "voices" that served as a type of conscience in determining what actions to take in his own life. These "voices" that Socrates refers to (Coppens 2012) influenced Socrates' decision not to enter into politics as a young man, despite his claims that "only philosophers were fit to govern the masses of the people." Socrates refers to these "voices" as "daemonic" and a function of his own soul. In today's language we might consider these voices as essentially our own "alter ego" or our moral conscience that guides our own behavior.

Thus, a historical analysis of early philosophy suggests that the first indication of the mind-body duality (i.e., thoughts influencing physical behaviors) begins actually through the works of Socrates and continues with the works of Rene Descartes and John Locke. Descartes has been given credit as having a profound influence in Western philosophy for having written several articles addressing the role of mathematics and using algebraic equations in describing geometric forms and shapes. He is also considered a rationalist and has been referred to as "The Father of Modern Philosophy" because of his emphasis on scientific interpretation of human behaviors as well as reluctance in embracing Catholic dogma in explaining human behaviors. He was considered a key figure in the Scientific Revolution (1550–1700) along with Sir Francis Bacon. However, despite Descartes' impressive scientific background, his revolutionary views were no match for the Pope and Catholicism and thus he abandoned four years of his work that explained natural phenomena via mathematical principles (*Treatise on the World*). This was especially important in light of the fact that Galileo was condemned and charged with heresy by the Catholic Church in 1633 due to his work in astronomy suggesting that the Earth is not the center of the universe.

Descartes argued in his groundbreaking work that all natural phenomena may be understood entirely through science and rational thought. Descartes also argued that all of nature could be understood through the principles of science and mathematics. Despite Descartes' strong emphasis on science and mathematics, he was wise to note God as the ultimate Creator and thus paramount to all of nature and causality. Because of his fundamental beliefs in math and science as well as religious beliefs, Descartes needed to be careful about how his views were published and perceived by the general public. He was very much aware of the censorship that the Vatican placed on any scientific theory that appeared to contradict those of religion, as evidenced through the ex-communication and house arrest of an earlier colleague, Galileo. Similar to earlier philosophers such as Socrates and Plato, Descartes argued that seeking truth and virtue in the physical world was a precursor to understanding wisdom. The most effective method in understanding wisdom was through an understanding of basic mathematics and physics, which later became known as the Cartesian Philosophy. Despite his loyal and profound beliefs in Catholicism, the majority of Descartes' works (i.e., *Meditationes de Prima Philosophia, Les Passions de Lame'*) were considered an affront to the

Catholic Church in Rome and in 1559 Pope Paul IV placed all of his works on the List of Prohibited Books. Rene Descartes was commissioned by Queen Christina of Sweden in 1650 to work as a private tutor, but unfortunately died of complications of pneumonia that same year. He was buried in initially in Stockholm, Sweden and later his remains were removed and interred in the Abbey of Saint Germain des Pres in Paris, France.

Perhaps some of Descartes' greatest contributions in philosophy and psychology included his work in geometry, calculus and physics. He should also be noted for his distinct efforts and ability to improve our understanding of human thought and reason through the development of mathematics and science in general. Much of Descartes' work in math and physics actually laid the groundwork for an equally impressive theoretician and scientist in completing works relative to physics and gravity—Sir Isaac Newton (1642–1727). Rene Descartes' work in mathematics and science was primarily an effort to empirically validate and investigate human behaviors relative to consciousness and awareness. He used his expertise in mathematics and science also to validate not only the existence of knowledge, but more importantly our ability to process and to comprehend knowledge to better adapt to our physical world. Descartes argued that while his mind is capable of producing knowledge, it is also capable of influencing and coordinating the physical movements of the body that have been associated with our thoughts. Additionally, despite Descartes' dedicated belief to Catholicism, he is credited with being one of the first major philosophers that challenged Catholic dogma through a scientific and mathematical analysis of natural events.

The Famous "Mind-Body" Problem

A very traditional (i.e., centuries old) problem that had plagued philosophers since the days of Socrates was that of the mind-body duality, or how does the mind purportedly influence physical movements of the body. This "mind-body" problem was one of the most significant topics of study that Rene Descartes pursued in his illustrious career. He did agree with earlier philosophies (i.e., Plato) that the mind indeed was separate from the body, but he further argued that the mind could physically influence how the reactions and movements of the body could be accurately measured. This theory of measuring reflex actions among human development was further investigated in his now famous work: "*undulatio reflexia*" which many consider to be the precursor to the works of Ivan Pavlov in stimulus-response, classical and operant conditioning.

Although Descartes argued that our minds have no physical or organic basis, he did note that the mind does control our cognitive or more commonly referred to "thinking" processes. When we "think" of something, such as food, this thought can actually trigger physical reactions, such as salivation and the salivary glands. Additionally, just as thoughts influence physiological responses, so does the application of physical stimuli (i.e., food) influence our cognitive and thinking processes. Thus, a major contribution of Descartes' approach to the mind-body duality problem was simply his emphasis on the reciprocal processes involving what the mind produces (i.e., our thoughts) and effects on our behavior (reflexes, such as salivation).

This important contribution soon became known as Descartes' Doctrine of Ideas. Descartes further argued that there were primarily two different types of ideas, those that are given to us by God (i.e., our capacities to comprehend abstract concepts such as God, infinity, moral justice and the "self"). These ideas were referred to as "Innate" ideas. The second form or type of ideas were more relative to materialistic events or experiences taking place on earth and included things that we "know" of by virtue of our experiences in the real or physical world. These ideas were referred to as "Derived" ideas and were a function of our experiences involving personal relationships, education, and communication.

In sum, few philosophers have had greater influence in psychology than Descartes. Descartes was the first philosopher to suggest that a reciprocal relationship of mind-body exists; prior to Descartes, philosophers argued primarily that the mind controlled dominant forces that primarily influenced human behavior. Other significant contributions by Descartes include the concept of "innate" ideas (such as God and infinity) as well as the idea that specific locations within the brain itself control aspects of behaviors (i.e., frontal lobe controlling speech) and the mechanical/muscular processes of the body.

Despite Descartes' emphasis on science and rational thought, much of his work describes the need to pursue (and fulfill) virtuous behavior in our relationships with others. Perhaps due to his close relationship with the Catholic Church, Descartes often discussed the importance of helping others in society. Through displacing our own needs with behaviors that helped others we can achieve grace and spirituality. Descartes argued that the need of the public should always supersede the needs of the self, and that the primary essence of human existence was achieved through spiritual sacrifice and displaying generosity to others that would improve community development as a whole.

JOHN LOCKE (1632–1704)

Mechanistic Principles of Behavior: Clock Work Precision

A common philosophical view during the mid-seventeenth century through the nineteenth century was that human behavior lacked free will and that our behaviors were controlled through natural laws and physics. This popular belief system came to be known as Mechanism and argued that all living things (organic matter) are related to each other and complex in nature. The existence and function of all living things are precise measurements, and that even our ideas are innate. Human behavior was thus considered to be a phenomenon that was predictable and controlled through laws of nature, similar to the speed of light, gravitational forces, etc.

The concept of Mechanism was first addressed by Aristotle and later discussed by Descartes regarding his views of duality and human nature. A common metaphor used in describing Mechanism was in comparing the functions of human behaviors with how a clock operates: Controlled, precise and exact. According to the Mechanistic theory, if we are capable of understanding the natural physical laws that govern our world, then we are capable of controlling and even predicting how human behaviors will unfold within the environment. Given the often destructive and violent behaviors that plagued Europe during this period of time, it is somewhat surprising that a commonly used metaphor in describing human behaviors was a clock or timepiece.

Rene Descartes introduced basic theories relative to rational thinking, such as mathematics, Cartesian Principles and is considered by many to be among the first of the modern philosophers. A contemporary of Descartes who would soon also make equally important contributions to philosophy and science was John Locke. As a young man, Locke found the contemporary academic topics (i.e., medicine) of 1647 at the Westminister School of London tedious and boring. Locke became more interested in the works of Rene Descartes and is considered to have made important contributions in the development of modern thought that has influenced related disciplines such as political and experimental philosophy, psychology, ethics, epistemology and modern liberalism. His publications are also considered to be one of the most influential pieces of the Enlightenment Era.

In 1674 Locke completed his bachelor of medicine degree and was consulted by an important political figure of that time, Lord Anthony Ashley Cooper (1st Earl of Shaftesbury). Cooper had invited Locke to stay with him for a short period and continue his work in medicine. Locke had helped Cooper in the treatment of his liver cyst in 1667, and medical procedure probably saved the life of Lord Cooper. It also provided an opportunity for Locke to develop some of his most important theories regarding human nature, government, and a theory described as the social contract: "Two Treatises of Civil Government" (1680). Locke's views in this paper treatise were highly critical of the absolute monarchy and control over human lives, and Locke argued that only God had absolute control over the events (physical as well as spiritual) that controlled the outcome of human life. The treatise basically was antithetical to the earlier works of Robert Filmer's (1588–1653) who defended the absolute control of the monarchy in his manuscript: *Patriarcha.*

Locke argued that due to the primary and inherent tendencies of greed among humans, laws must be established in society to govern and control fairly the rights and property of others. Locke recognized that cooperative relationships within smaller groups were clearly possible (as discussed in the Social Contract Theory). He was also aware that human nature was capable of opportunistic and egoistic behaviors that could threaten basic principles of fairness and democracy within society. In this sense, then, Locke agreed with Thomas Hobbes to allow communities to develop in terms of commerce and business, but that some form of rule or law would also be necessary in respecting individual rights.

In this way, Locke's political views were very similar to a contemporary of his, Jean-Jacques Rousseau (discussed later in this chapter) (1712–1778) in that political forces and governments are necessary developments in human civilization in light of human greed. However, with the development of a government for groups of individuals we need to remember that individual rights must be surrendered to protect the civil rights of all persons. It was also during this time that Locke began formulating his views regarding the plasticity of human nature and human behaviors. No single individual is born with any type of "predisposition" or tendency that influences their behavior to be prosocial and cooperative or antisocial and destructive. According to Rousseau and Locke, all human behaviors reflect the types of environmental experiences that an individual is exposed to. Locke further argued that four primary factors contributed to all learning: Repetition, Association, Rewards and Punishment and Imitation. These four primary constituents of behavior are strikingly similar to the fundamental concepts of modern Behaviorism, that argues all consequences of behaviors (i.e., positive effects of a particular behavior tend to be repeated; negative consequences tend to be withdrawn or extinguished).

JEAN-JACQUES ROUSSEAU (1712–1778)

The work of Jean-Jacques Rousseau was timely in the sense that his work influenced the development of educational and political views, as well as the French Revolution and eighteenth-century Romanticism. Rousseau was a champion of human rights within society. He believed that not only can humans engage in a productive and cooperative society with each other but communities were more efficient and productive when individuals are provided with opportunities to develop their skills and share resources with each other. His views regarding human nature, civilization and compassion were unique in the sense that he disagreed with earlier philosophers such as Thomas Hobbes arguing that while all humans do possess the capacity of destruction based on greed (i.e., what some have referred to as the "savage state"), Rousseau argues that human civilization does in fact operate by principle and that humans aspire to be the "noble savage" in achieving social harmony. A common metaphor during this time was the comparison of human behaviors with a caged animal; Humans can display cooperation and compassion when guided by their moral principles as well as the structured laws of society. However, Rousseau also argues that is would be naïve (if not irresponsible) to forget the depths of depravity in which humans have engaged in to maximize pleasure or profit:

> *The example of savages, almost all of whom remain it always . . . this is the veritable youth of the world; and that all the subsequent progress has been in appearance so many steps toward the perfection of the individual . . . and in fact the decay of the species.* (Rousseau 1754)

Rousseau argued that the "passions of the soul" (i.e., sexual urges) were part of our basic and primitive nature, and that the society and communities in which people live must recognize the "brutish and savage" nature of all persons. However, Rousseau also identifies the positive and cooperative side of human nature, and that the capacity to live in social harmony with each other is not only possible, but necessary for continued existence. Rousseau also disagreed with Hobbes in the sense that even though our basic and primordial nature is selfish behavior, humans do retain the capacity to live in a morally "uncorrupted" manner and distribute our skills in such a way to promote harmony among all people. When provided with opportunities of self-advancement and professional growth such as community service projects, Rousseau argues that the natural tendency for all people is in positive growth and community development, without a need for enslavement or punishment:

> *In any case, frequent punishments are a sign of weakness or slackness in the government. There is no man so bad that he cannot be made good for something. No man should be put to death, even as an example, if he can be left to live without danger to society.* (The Social Contract 1762)

The fact that Rousseau argues that all individuals are capable of making positive contributions to society and culture represents his belief in human potential despite our primitive tendencies of greed and selfish acts as described previously by Hobbes (Curran 2002). We are capable of moving from the "savage state" to a more composed "noble savage" as described by our actions within community and social engagement. This statement by Rousseau clearly states his belief in the redemptive nature of human beings;

He also argues why societies may benefit when providing persons with opportunities for personal and professional development. A reciprocal relationship (now referred to as social capital) develops where both individuals and communities flourish with opportunities for growth and development. Rousseau did not believe in inherently evil persons or behaviors; according to him, only corrupt societies corrupt human nature.

The existence of a civilized society is not only possible, but a necessary function of our own evolutionary history that provides individuals with opportunities to exhibit and distribute their trade and skills with each other. Rousseau argues in *The Social Contract* (1762) that democracy is only possible when individuals realize that their "human rights" do not supersede the rights of others, and that the primitive human nature that we all share is highly vulnerable without the existence of sovereign laws to govern and guide our behaviors in a fair and equitable manner.

One of the most famous quotations from Rousseau: "All men are born free . . . but we live our lives in chains" refers to the fact that we typically accept arbitrary rules without really examining how these social rules and regulations may impact our lives. Rousseau argues that because of human nature we need to be cognizant of the fact that some will inherently attempt to exploit others based on greed. Ethics, social responsibility and moral values are in fact achievable constructs within our society, but only through conscious and organized deliberation by each member of society. The analogy or metaphor of the lone or solitary animal is often attributable to the philosophical works of Rousseau because they reflect the duplicitous nature of humanity; even vicious animals can live in harmony if certain needs are met. Because humans may be predisposed to behave in an uncivilized manner this does not mean that this is our destiny.

People who are living within the community or society *need* a democratic process to protect them from a basic and primitive greed found universally within all humans. Rousseau argued that a form of government that is democratically elected to fairly monitor and control a variety of human interactions, including commerce, business, and distribution of property was vital to civilization. Without it, Rousseau insists that anarchy and social chaos would be inevitable. Thus, it is important to note that Rousseau does not claim that it is *impossible* for humans to coexist in a meaningful and civil manner. On the contrary, Rousseau argues that the only way humans can coexist (commercially or interpersonally) is through our submission and recognition of some form of government that all community and society members recognize and obey. Rousseau would also agree that the more productive communities are those that emphasize the mutual benefits of cooperative behaviors (thereby eliminating the need for a sovereign power) and eliminating those economies and environments that emphasize a "zero-sum" relationship (one's gain comes only at the expense or cost of another person) relationship among group members.

As people become increasingly dependent on each other in society, there is also an increasingly important need for people to recognize and respect the rights of others to maintain and preserve civilization. We become aware of our own skills and traits as a means of sharing them in commerce and community development, this being a positive trait that promotes development and interdependency with others. However, Rousseau also identifies the common human tendency for a type of "self-love" or avarice to develop in community or social settings that results in self-pride.

Humans, from an evolutionary perspective have capacities for both cooperation and egoistic behaviors; it is thus a requirement for individuals to recognize the need for community regulation and laws that exist as safeguards of our basic human rights. By the development of our skills and trades we create a society that allows individuals to live in a fair and equitable society that represents the will of the people or democracy. Rousseau argues that the laws of our society not only allow individuals to live and progress in a healthy manner, but that the laws of the community are necessary to protect us also from our aggressive and "savage" nature that exists in each and every one of us. These views are further elaborated in Rousseau's "*The Social Contract*."

Perhaps the most impressive component of Rousseau's "*The Social Contract*" is his argument for individual freedoms, civil rights, and basic human equalities that we all can and *should* share. In the beginning statement of The Social Contract, Rousseau argues that the oldest and most ancient of all societies is that of the family. Healthiest of all societies (and families) is the union of interdependency and community—a mutual sense of respect, sharing and contributing for both ourselves and those we share goods with. Similarly, Rousseau takes issue with both Hobbes and Aristotle regarding human nature, human rights and the topic of slavery. Aristotle argues that some individuals (by some fatalistic perception of their own destiny) have been predestined for slavery and imprisonment. Eventually those who are imprisoned even lose their own drive and desire for their God given right of freedom and thus sadly become institutionalized. Rousseau argues that no man is "born" a slave, and that the terms "slavery" and "human rights" are inherently contradictory. In another famous quotation from "*The Social Contract*":

> *Every man having been born free and master of himself, no one else may under any pretext whatever subject him without his consent. . . . To assert that the son of a slave is born a slave is to asset that he is not born a man.* (p. 4)

All persons have inherent rights of freedom, but we sacrifice those rights when we relinquish ourselves to the arbitrary nature of laws governing society in an unfair way with an unequal distribution of resources and social capital.

The early Greek philosophers such as Socrates and Plato argued that the human self-knowledge first begins through introspection and critical self-inquiry. The critical search for truth through self-interrogation was the primary way to achieve happiness not only for oneself but also critical in achieving community and social harmony. The world and community itself can only become virtuous and valuable through a process of self-inquiry and examination of one's own behaviors. When individual citizens take greater responsibility for their own actions, then the welfare of society and community as a whole exponentially improves. Socrates and Plato both argued that knowledge as an absolute would be impossible to attain; however, wisdom is acquired through daily questioning of things that we take for granted. When individuals regularly participate in the questioning of their own behaviors through self-analysis and contemplation, the potential for social justice emerges through self-awareness (Simon 1973), whereas conflict and chaos is reduced within the society. As early as 380 BCE we see through the works of Socrates and Plato that a just and virtuous society is not only considered an abstract ideal but something that can be determined through self-inquiry and introspection. When each group or "class" of individual corresponds to reason through self-analysis, communities and nations may be controlled through a democratic process of social justice and harmony.

Helping Behaviors and Reciprocal Altruism

The evolutionary benefit is that communities become inherently stronger and more resilient when groups work cooperatively together, thereby increasing survival among all members within the community itself. Humans are not the only organisms that have displayed the qualities of reciprocal altruism, as the lower primates (i.e., baboons and chimps) and even birds, bats and insects have been noted to engage in the practices of reciprocal altruism (Wilkinson 1984).

Additionally, when individuals within groups engage in principles of reciprocal altruism, the entire group itself becomes significantly stronger and more resilient. The single most effective strategy in winning a zero-sum game situation (i.e., Prisoner's Dilemma) is simply replicating and matching your partner's responses. A "tit-for-tat" response means that people do not exploit their opponent (even when it seems advantageous to do so) and maintaining a "nice" playing strategy will significantly influence future potential partners to play similarly in a "nice" or cooperative manner.

Individuals who display altruistic behavior to a nonrelative within a community or group are more likely to receive similar help in future situations and the group as a whole has benefitted. According to the philosopher Thomas Hobbes, a major component of human nature is that of greed and selfish behaviors. Egoistic behaviors have existed as long as humans have existed, and if society and communities can exist at all, there must be some form of relatively equal exchange of favors between people within groups. If humans are to be saved from their own greed, destruction and anarchy, then a sovereign ruling force must exist. Cooperative behaviors can exist but only under the threat of sanctions if individuals violate the norms and laws of a society. Competition within a capitalistic society drives both cooperation and distrust among community members, and society can only exist if people are aware of what humans are capable of doing to each other. Successful cooperative (either individual or group) relationships operate on three simple and mutually beneficial principles: Trust, cooperation and reciprocation (see Figure 2.2 below).

BUILDING COOPERATIVE COMMUNITIES

a. Take the Initiative—Positive First Steps: Don't wait for someone else to start a new program that helps communities—why not take the first step yourself? Help establish programs that facilitate interaction and development among community members and support groups (neighborhood watch programs, "if you see something, say something", volunteer for schools, etc.;
b. Don't become the "Bad Guy": Avoid opportunistic situations that provide benefits to you but may cost the community (i.e., initial defector);
c. Tit-for-Tat: Play fairly and equivocally with others within the community:

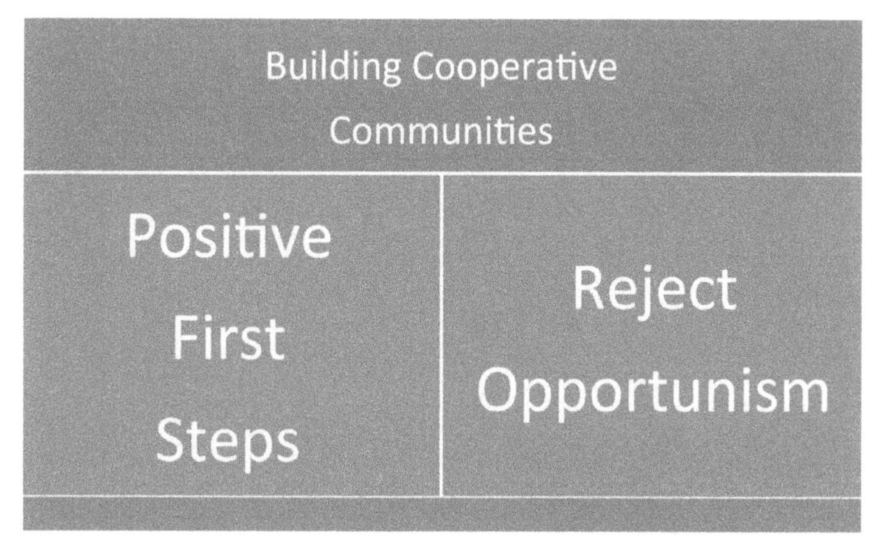

Figure 2.2 Building Cooperative Communities.

Chapter 3

Structuralism and Functionalism

BEGINNINGS OF MODERN PSYCHOLOGY

Beginning Schools of Thought in Modern Psychology: How Structuralism Influenced Evolutionary Psychology

An interesting note among scientists who have made brilliant contributions to the field of science is simply how misunderstood they were as children growing up. Some of the greatest thinkers, philosophers and scientists today were unfortunately labeled as "slow learners" or even "learning handicapped" due to their different or atypical styles of learning and their difficulties in establishing social relationships. Albert Einstein, for example, was considered "mathematically inept" through primary and secondary grades due to his perceived inattention in class. Wilhelm Wundt was also perceived to be "dim witted" and a slow learner throughout his work in the German Gymnasium. He was perceived to be socially awkward and even "antisocial" among his peers, and it wasn't until completing his doctoral work at the University of Heidelberg (1855) that Wundt was perceived as having unique contributions in the discipline of psychology (Schultz and Schultz 2012).

Considered by many historians to be the first modern discipline in psychology, Structuralism attempted to understand human behaviors from a consciousness perspective and that interpretations of reality are essentially based on experiences involving physical energy within the environment. The Structuralist attempted to understand the human psyche by analyzing human consciousness into separate components and view the mind through a process referred to as introspection. Additionally, consciousness and awareness was influenced by the environment at three primary levels: Sensations (physical energy perceived and transduced from the five basic senses); images (mental representations that exist as a function and consequence of experience within the environment); and finally affect (what we typically characterize as "mood states" that exist as a consequence of experiences in our relationships with others within our environment).

The early Structural psychologists such as Wundt and Titchener helped set the stage for evolutionary psychology to develop by their emphasis on sensory adaptation and environmental experiences as key elements that helped humans to adapt to their physical environment. Evolutionary psychology explores how behavioral and psychological traits have developed over long periods of time through a process of natural selection. Those behaviors that proved to be highly adaptive and suitable for survival were most likely to be retained and passed on to future generations. Humans that were more likely to be consciously aware of important characteristics within their physical environment that could threaten their own safety and reproductive fitness (i.e., predators) could have a clear advantage to survival and protection of offspring.

Wilhelm Wundt (1832–1920)

Wilhelm Wundt is considered to be the "founder" of modern psychology because of his unrelenting work ethic and productivity. He taught and lectured at the University of Leipzig for over 45 years, founded psychological journals (i.e., *Psychological Studies*), served as both reviewer and editor for several journals, and published (1873–1874) what many theorists consider to be one of the most influential manuscripts in psychology: "*Principles of Physiological Psychology.*" Wundt is also credited with discovering and developing a controlled research environment in psychology (i.e., the laboratory) where topics of research such as consciousness, sensation and perception could most accurately be studied. In a word, Wundt is credited essentially as a founder of psychology because he was the first individual to pursue and study psychology as a *measurable and controlled* science.

Wilhelm Wundt and Edward Titchener first described Structuralism as a branch of psychology that primarily addressed the role of introspection and self-analysis as the best method of understanding behavior. According to these theorists, psychology's main role in addressing and understanding human behavior was through analysis of consciousness, and that these units of consciousness were the cornerstone of our awareness and existence. An important function of the Structuralists was in defining how physical elements in the natural world influence our states of consciousness and awareness, and ultimately shaping and defining perceptions of knowledge. Interestingly, the arguments presented by the Structuralists in defining objective reality sound very similar to the early theories of Rene Descartes in his attempt to quantify knowledge and experience in a rational manner. Behavior and knowledge were considered to be a function of one source: Experience. The Structuralists were not so much concerned with what the mind *did* in terms of functions and purposes (this was addressed by the Functionalists), but rather they were concerned with the *mind itself*; What was it actually composed of? How does the mind actually process sensory information and how does this sensory information become translated into meaningful information?

The Structuralists often explained human behaviors in terms of basic units of experience. These basic units were often compared with atoms—molecules that could not be broken down into smaller components. They were seen as tools as a means of providing a "stepping stone" to a better understanding of our world and our environment. The Structuralists argued that before we can understand the functions of our behaviors we need to first understand the structural development of the elements that make up our experiences within the environment. The Structuralists (i.e., Wundt and Titchener) were not so much concerned with what the mind could *do* so much as what the mind actually *was*. Too often, according to the Structuralists people confuse perception with individual elements of reality and matter. Instead of saying we see a round red object with a stem at the top, we claim to see an "apple." This common perceptual mistake is what the Structuralists referred to as stimulus error: Confusing the sensory experience of elements with their perceived meaning. Consequently, humans must first look inward through our own experiences and better understand these units of consciousness and not confuse them with culturally defined interpretations of meaning.

Understanding the relationship between physiological and sensory influences was critical to understanding the basic tenets of Structuralism. After two publications: "*Contributions to the Theory of Sensory Perception*" (1858) and "*Elements of Psychophysics*" (1860), and later a third textbook publication: "*The Principles of Physiological Psychology*" (1867), Wundt was creating an environment of scientific exploration through the development of experimental laboratories as well as introduced the foundation of his work addressing the relationship between sensory perception, units of consciousness and introspection. The year 1875 proved to be an important milestone in Wundt's professional career. For the next 45 years he devoted his life to research and teaching at the University of Leipzig. He is credited as being the first psychologist to utilize the laboratory in his research and to establish the scientific method in psychology in Leipzig, Germany. These years would prove to be pivotal in Wundt's ability to introduce what many scientists describe as the first modern branch of psychology—Structuralism. For many historians in psychology, Wundt's influence in the development of modern principles of in psychology remains unprecedented (Blumenthal 1980).

The scientific climate during the seventeenth and eighteenth centuries in Western Europe was rapidly changing and was comprised of individuals who tried to describe human behaviors more from an internal (i.e., introspective) analysis. Prior to the introduction of functionalism, understanding human behavior was generally regarded as an individual, metaphysical and psychological experience. Furthermore, the Structuralists argued that behaviors were primarily a function of psychological interpretations that were based on physical experiences within the environment. Wundt combined educational theory, philosophy and biology and is considered by many to be one of the most prolific late-eighteenth-century thinkers. His views shaped how modern psychology evolved into more contemporary disciplines in psychology such as functionalism and Gestalt psychology.

What made Wundt's work unique was his very precise method of observation and controlled experimentation. He did something rather unique relative to scientific methodology, and that was to very meticulously record his own reactions and responses to a variety of stimuli. Wundt required that his laboratory assistants practice recording their observations when training in introspection over 10,000 trials of recorded visual and auditory sensations. Additionally, because of his meticulous procedures in laboratory studies and methodology, Wundt's work established basic guidelines and standards in experimental procedures. These practices have established principles that are still used today in laboratory procedures ensuring methodological reliability and validity.

Wundt and his student Edward Titchener conducted several now famous experiments in his new laboratory in Leipzig, Germany. Perhaps his most famous experiment explored perception and environmental influences of the sound of a ball hitting an object. Interestingly, Wundt discovered through his early research that the degree of the complexity of a given task requires slightly longer processing times when hitting a key in the laboratory. More importantly, Wundt was able to record most of his scientific findings combined with his views addressing human behavior in what remains to be a significant contribution to psychology: *Principles of Physiological Psychology* which was published in 1874. Wundt was very instrumental in shaping the future of modern psychology, and his work at the University of Leipzig in Germany influenced numerous American psychologists, including William James, Titchener, G. Stanley Hall, and James Rowland Angell.

Human consciousness was then (as it still remains today) a very complex concept to rationally describe and understand. The Structuralists argued that all behaviors (and thoughts) can be broken down to elemental units of consciousness. Behaviors and even conscious experiences may be described as basic types or units of elements that define our life experience. The Structuralists were known as a group of psychologists that argued that since all behaviors are based on unique subjective experiences, the best way to understand behavior was through controlled introspection, a belief system that is similar to the early philosopher Plato's famous maxim: "Know Thyself."

According to the Structuralists, the best method of understanding one's behaviors was simply a matter of critical self-analysis and looking inward via introspection as a means of better understanding your behaviors. Knowledge and insight were contained "within" but extracting the information was a more complex matter.

Plato and the Greeks argued that knowledge only comes through contemplation and people experience the physical world on three separate or different levels: The "Ideal World" which is best described as the world and environment that surrounds us; the "Material World" that was developed by a spiritual form (Deity) and the "Psychological World" which describes how individuals perceive their world based on their experiences in the world. It is here where we see how differences in the perceptions of reality influence different psychological states. An individual who believes that he or she controls their own destiny and has a strong internal locus of control has probably had extensive experiences in controlling the outcomes of their actions. Conversely, individuals who have an external locus of control have probably had minimal success in controlling successful outcomes of their behaviors and thus feel limited in their ability to modify their own behaviors.

Throughout history psychologists have argued between the existence of theory and reality which seems to resemble very closely the controversy the Greeks encountered in the descriptions of human behaviors. Emotions, such as happiness, sadness, desire and lust were considered very "mortal" and destructive to scientific reasoning and logic. Arguably, even today there are many theories that challenge the utilitarian needs of emotions. Freud, for example, argued that emotions were a function of repressed instinctual desires. Self-control, rational thought and logic were all attributes that were considered fundamentally necessary within science.

Abnormal Behaviors and Biology

Modern theory in psychology now views organic factors and biology as critical components to better understand human behaviors. The brain, central nervous system and neurotransmitters now explain how our physical bodies respond to changes in neurological and hormonal functioning. The early Greeks again seemed to understand the critical role between physical and psychological functioning. Early philosophers such as Hippocrates (460–377 BC) and Galen (129–161 AD) argued that physical and psychological illnesses must have natural causes, and that the best understanding of human behaviors involves a rational approach ("*That the Best Physician is also a Philosopher*", Claudii Galeni Pergameni 1992). These natural causes were a result of "imbalances" of critical fluids that he characterized as yellow bile, black bile, blood and phlegm. Interestingly, Hippocrates also noted that personality characteristics or types may be predisposed to specific types of ailments. Psychological ailments such as depression or anxiety were considered to be related to organic or physical problems such as fluids or the central nervous system. This information initially described by Hippocrates during the fourth century was fundamental to improving our understanding of the relationship between psychological and physiological states and conditions.

Interestingly, current theory today still supports the observations made by both Galen and Hippocrates. We know, for example, that coronary heart disease is significantly more likely to develop among those individuals who have been characterized as "hostile and competitive" or "Type A" (Friedman and Booth-Kewley 1987). More recently "Type D" has been described as an overall negative affect characterized by irritability, worry and anxiety (Denollet, Martens, Conraads, and de Gelder 2008). As psychology continued its growth in Europe, there was an increasingly stronger emphasis that it should in fact be a science that was capable of replication and observation. Wundt's student, Edward Titchener continued his work addressing the relationship between self-reports of experiences within the environment and comparisons with events in the physical world. As a consequence, concepts such as introspection, self-analysis and reports of the events in the physical world became the foundation of the scientific study of structuralism.

Interestingly, events that triggered further study in structuralism was the common phenomena of groups of individuals experiencing a common physical event in our world, but reporting vastly different accounts. Titchener argued that while it may be true that singular events do occur within our world, the fact that individuals often report different events suggest that we process this information differently based on separate experiences within our environment. The limitations of structuralism, however, were the subjective reports people provided and the uncertainty of knowing what factors contributed to specific emotions and behaviors.

When individuals discuss the topics of "Cultural Psychology" today, few would consider the contributions of the early European psychologists. However, it actually was Wilhelm Wundt who argued that in order to accurately understand psychology as a science involving topics such as memory, sensory perception and learning, psychologists must first consider also culture in which individuals have been exposed to. It was Wundt who first argued in his ten-volume text: "*Cultural Psychology*" (1900–1920) that human mental and psychological development was first influenced by society's mores, values and accepted cultural practices. Indeed, Wundt also is credited with some of the first publications addressing groups of individuals interacting with each other in communities—a topic today that is better known as social psychology.

The Role of Consciousness in Structuralism

Perhaps one of the most important components of Structuralism was that of consciousness and conscious experiences as playing a pivotal role in our behaviors and interactions with other persons within our community. Empiricism and Structuralism share few similarities with the exception that elements are learned and processed by individual experiences within our environment. Thus, Wundt would argue that empirical knowledge differs not because of differences in actual elements, but rather how individual *processes* each event. Perceptions of events may be more influenced based on the unique experiences within those environments, thereby explaining differences in human adaptation and behaviors based on each event. Furthermore, individuals who may be exposed to specific stimuli (i.e., resources such as edible plants and animal species) within a particular geographic region may have had distinct advantages (i.e., hunting skills) compared to others who lacked similar experiences to stimuli (Tooby and DeVore 1987).

As a matter of survival humans living in groups had to cooperate in specific ways as a means to adapt to a variety of different types of environments. This meant understanding one's environment in such a way that these specific necessary resources (i.e., food) could be identified and harvested as well as avoiding conflict with potential predators. Evolved mechanisms of understanding how to survive and adapt to resources specifically in one's environment has also been referred to by several evolutionary psychologists as a form of mediated environmental science, or more commonly referred to as "folk biology" (Atran 1998). The concept of folk biology is an important universal theory that describes the how the characteristics of various resources and elements within the environment can facilitate adaptation and pose as a defense against predators. This intuitive form of knowledge can be shared by all members of a species (i.e., humans having an evolved preference for specific types of sweet and high caloric foods) and has been credited by many evolutionary psychologists as a key mechanism in human adaptation (Buss 2015).

Wundt argued that despite individuals recognizing events and stimuli within their environment as being similar events, our perceptions of these events differ not because of actual differences within each element but rather due to our perception of each event. An important related concept that Wundt also addresses is voluntarism, or individual capacities to differentiate mental components through higher-order thought processes. Knowledge and higher education is based on the preliminary concepts of voluntarism, or our ability to organize information from basic levels to more advanced or complex models into higher-order cognitive processing systems.

Introspection and Structuralism

Imagine the last time you experienced something very positive and favorable—a trip to a favorite vacation spot or simply visiting your friends. Think back about what it was specifically that made those events positive and happy—was it a conversation, the environment, or was it simply sharing experiences with your friends? The ability to look inward and to reflect about past experiences can also shape our perception of the future. This rather unique human ability is referred to as *introspection*. The early Structuralists emphasized the role of consciousness and awareness as critical features of understanding our own behaviors. According to Wilhelm Wundt, a critical component to understand our behavior was in our ability to analyze and organize our experiences and perceptions that are inherently unique to each individual. Thus, according to both Wilhelm Wundt and his successor Edward Titchener, how individuals perceive and understand events within their environment is subject to variation because of different experiences from each person. An example might include eating your favorite flavor of ice-cream, such as vanilla. Several people may be experiencing the same flavor (vanilla) but their individual reports of experiencing this flavor may vary considerably. The flavor itself is not different, but rather how people experience the sensation of the flavor itself is what is different. Other sensations may be explained in similar ways by the Structuralists, such as light, sound, touch, and smell. We may agree that a singular and consistent stimulus exists, but it can also be expected that individual reports of these perceptions and stimuli will vary significantly.

Titchener and Wundt both argued that introspection was a critical component of the process of understanding mechanical events in the physical world, but participants needed to be individually trained in this process. An important component of this process was first in teaching individuals directly what they were experiencing in terms of sensation rather than reporting a simple reaction. Participants in the introspective process were described as a mechanical process that included routine and habitual behaviors. Consistent in the theory of introspection was the description stimuli containing similar properties. These properties included terms such as "duration," "clearness," and "intensity."

The process of introspection was fundamental to the theory of Structuralism, however, it soon became criticized by many leading scientists in the early eighteenth century. The primary source of criticism came from the subjectivity of the reports given by individuals who may have been experiencing stimuli and sensations within their environment. Many theorists, such as August Comte, have argued that there are physical limitations regarding how the mind views itself and if it is possible that the mind observe and monitor its own activities. Is it possible, for example, to understand the emotion "happiness" without actually experiencing it? Is it possible to understand the emotion "depression" without ever feeling sadness or remorse? As you can see these are fundamentally important philosophical issues that still plague the followers of Structuralism even today.

Wundt attempted to scientifically analyze all physical properties (i.e., sounds, visual stimuli) and quantify their existence relative to conscious experiences. While different people may process physical stimuli (i.e., sounds and visual imagery) differently, Wundt argued that if psychology was in fact to be defined as a science, classification of different stimuli must be identified and described. Wundt emphasized that experiences were the key elements to understanding human consciousness, and that consciousness could best understand the "outside" or physical world through two different types of subjective experiences: "Sensations" and "feelings."

Wundt attempted to explain how individuals report their subjective experiences through consciousness. "Feelings" were described as subjective emotional consequences of how we experience events in our work as well as physiological stimulation. Visually experiencing your child play may result in a report of *feeling* pleasure, caused by the visual physical stimulation of impulses sent to the occipital lobe of the brain.

Wundt's Tridimensional Theory of Emotions

How individuals physiologically experience events within their levels of consciousness has been defined by Wundt as the now famous "tridimensional" theory of subjective experiences: Feeling states are defined at three mutually exclusive or different levels: positive/negative; tension/calmness; and excitement/depression. Wundt also argued that when individuals perceive events within their world, they tend to not view elements in a separate or individual way, but tend to organize elements into a more meaningful or "whole" process. This tendency to organize individual or separate elements into meaningful combined portions of our experiences has been described in great length and has even contributed to a new discipline in psychology: Gestalt psychology. This newer German discipline of psychology was later expanded by psychologists Max Wertheimer, Kurt Koffka and Wolfgang Kohler.

According to Wundt, how individuals experience events within their environment can tell us how they process conscious events. Wundt described a new basic tendency to process events in a way that they provide meaning to the individual viewer. In other words, when we view a common vegetable such as a red tomato, we do not identify the stimulus as a round, red vegetable with a green stem, but rather we view all of the characteristics of the stimulus simultaneously to help us process and identify the stimulus as a tomato. Thus, our "mediate" experiences are those visual elements that we first process to provide meaning to our world. Conversely, "immediate" experiences are those initial visual experiences that we perceive first without subjective bias or interpretation.

Apperception: Organizing Physical Stimuli into Meaningful Elements

As Wundt labored in defining his theories as they related to conscious experiences, other theorists were criticizing the basic concepts of Structuralism as essentially an oversimplification of more complex physical structures. The Gestaltists (Max Wertheimer and Kurt Koffka) argued that the principles of Structuralism—perceptions of reality based on conscious experiences of smaller elements built upon larger elements, was simply a "stone and mortar" approach. The "stones" according to the critics of Structuralism simply comprised the physical events of the world and the "mortar" referred to the associations that we have with each experience. The fundamental experiences that we all share can only be comprehended through the process of what Wundt referred to as "apperception." Apperception was described as the process by which mental elements (i.e., thoughts or ideas) are organized into meaningful experiences, a term that he used earlier in referring to "creative synthesis." While Wundt described the process of apperception as a means of furthering the understanding and comprehension of Structuralism, the Gestaltists soon began to provide a different perspective of conscious experiences and how these experiences shape our perception of reality.

The Gestaltists argued that individualized or "atomic" experiences are separate from our perception of reality. When we try to define objects in our world, we first combine sensory elements of stimulation (such as color, density, depth, etc.) to help us understand what the stimulus is. A round orange object, for example, is described as an "orange"—We do not first describe an orange as a round rind with edible fruit. We first combine these visual physical characteristics (orange) and identify it as an edible fruit. The now famous phrase often used by Gestaltists: "The whole is more than the sum of the parts" simply means that the things that we typically experience in our physical world occur in one organized event. When we see a tree, for example, we do not see a grouping of several thousand leaves, branches and twigs, but simply a tree that is comprised of those elements.

Wundt is considered by many to be the founder of modern psychology because he *intended* the early discipline of psychology to become a dominant science. Arguably other theorists made important contributions in psychology (some even earlier, such as Fechner), but those theorists were simply contributing to psychology via experimentation and research publications. Wundt not only was an ardent scientist who frequently published journal articles (he typically published over 2 articles a year), his intention was to create psychology as a leading scientific discipline. His work was also competing with several other important contributions of other scientists in the late nineteenth century, notably Charles Darwin and William James.

As Wundt continued his arduous work in the laboratories of Leipzig, Germany, he and his assistant Ludwig Lange attempt to correct stubborn discrepancies in recording reaction times to varieties of stimuli applied to subjects (i.e., withdrawing hand from an open flame). Wundt refused to even consider the increasingly popular theory of evolution to explain individual differences in reaction times or the influences of environmental consequences on these internal events. Wundt insisted on the universal rule that human reaction times were in fact consistent across cultures and geographic locations: however, his results were inconsistent. J. M. Baldwin (1896) coined the term "organic selection" to account for the variability in human reaction times, and further argued that the evolutionary theory that Darwin proposed earlier in 1859 would actually *support* the variation of reaction times based on differing biological adaptive characteristics that were common in different geographic locations. The differences in biological adaptations in reaction times would be transmitted to future generations, thereby providing strong support for Darwin's theory of evolution and natural selection.

Shortly after the publication of Darwin's *On the Origin of Species* in 1859, many scholars began taking a much closer look at the relationship between the environment and behavior. Topics that were once popular in Germany addressing internal components and mental structures influencing consciousness were now being replaced with topics addressing how organisms function and adapt within their environment. It seems fitting that the name of this new science in psychology would be Functionalism, and that a major proponent of this science was Darwin himself.

As Wundt's research program eventually became stagnant during the late nineteenth century, research in the United States with William James and Functionalism was just beginning to flourish. It seemed that a whole new era of scientific and psychological discoveries that were prompted by evolutionary theory were about to unfold in the United States. Recent research that explores the relationship between Darwin's theory of evolution and the psychological discipline of Functionalism now describes Darwin's work as the primary driving force in the development of the American branch of psychology that we refer to as Functionalism (Green 2009).

FUNCTIONALISM

Often considered to be the founder of modern psychology in the United States, William James is typically considered to be a central force and influence in the introduction of Functionalism as well as identifying psychology as a science in the late nineteenth and early twentieth century. Despite publishing one of the most significant textbooks in psychology (*The Principles of Psychology*) in 1890, James did not even consider himself a psychologist and often criticized the very discipline that he helped create by stating that "psychology is merely an elaboration of the obvious" (Schultz and Schultz, p. 131). Furthermore, while James focused on the importance of consciousness in understanding the physical elements of the environment, he was derided by colleagues because of his interest in abstract concepts such as parapsychology, clairvoyance and spirituality. His life was prone to several bouts of depression, anxiety and what was then referred to as "neurasthenia" (or nervous tension). Despite his numerous physical ailments, James has been regarded as one of the primary influences in the development of modern psychology and is noted as being the first educator to offer undergraduate psychology courses at Harvard University in 1875.

William James has contributed to several areas of philosophy that have also had an important impact in the development of modern psychology. James argued that a distinct human trait and character is that of our own awareness of the self and consciousness. Humans have the distinct capacity to recognize and understand that perceptions of awareness and causality may be inherently distinct among different groups of individuals. William James' approach to rationalism and pragmatism is such that humans are capable of understanding abstract concepts such as the meaning of life without relying on spiritual or religious truths. Any human activity that provides security and adaptation to our environment is therefore useful to our own survival, and this does include religion.

Functionalism was becoming a very popular branch of psychology in the eighteenth century because it helped explain human behavior more from science rather than from theory. Functionalism adapted and changed with the new zeitgeist of the United States—principles of industrial and organizational psychology were now used in public schools (i.e., testing), clinicians were treating clients with a variety of disorders, and businesses were also using psychological theories to maximize production and sales (Landy 1997). The publication of Darwin's *On the Origin of Species* (1858) had a profound impact in all disciplines of science, notwithstanding psychology. William James became directly influenced through the works of Charles Darwin and described Functionalism as a science of what the mind could do in relation to experience within the environment. Functionalism takes into account the role and influence of environmental factors and adaptive behaviors as a more *reciprocal* relationship. The focus of Functionalism was in understanding how the brain adapts and functions within the environment and does not emphasize how the brain is constructed or structured in processing consciousness.

The Functionalists took a radically different approach to understanding human behaviors. They argued that the key to understanding human behaviors is not through the analysis of our own thoughts (i.e., introspection), but rather through an understanding of our own *mental processes* through experiences and interactions within our environment. The Functionalists were more concerned with what the mind actually *did* relative to the physical environment rather than the physical and structural makeup of the mind (a primary interest of the Structuralists). Functionalism was therefore considered an effect or outcome of one's direct experiences within the physical environment. The distinction between Darwin's classic theories addressing evolution and the process of natural selection and Functionalism was that Darwin focused on the *physiological* (i.e., phenotypical) changes that occurred over time through the process of natural selection, whereas the Functionalists were concerned with our cognitive development (intelligence) through an interaction with the environment.

According to William James and the early Functionalists, behaviors (along with instincts) needed to serve an adaptive function that enabled people to better understand and master their environment. Behaviors became patterned and routine as they served a variety of adaptive purposes. Smaller groups (i.e., clans) that displayed key (intragroup) social behaviors such as trust, altruism and cooperation were better equipped to survive harsh environments, predators as well as aggressive outside clans (Yamamoto and Takimoto 2012). Those behaviors that produced favorable or positive outcomes tended to repeat themselves within a variety of different types of environments. As behaviors became more predictable, William James argued that consequences within the environment shaped and help develop important human traits that later were referred to as "personality" traits or characteristics.

Human behaviors must serve an adaptive role in relationship to the environment in which an individual was exposed to. Thus, changing or differing environmental topography would ultimately account for differences not only in behavior but cognitive and intellectual development. Scientific knowledge (i.e., empiricism) was defined in a way that enabled researchers to replicate findings in a controlled manner, which is still considered to be an important component of research validity and reliability. This information was recorded in a systematic and empirical manner based on scientifically recorded measurements.

William James has offered many fascinating contributions to the disciplines of both philosophy and psychology, and is arguably considered to be one of the most influential philosophers and theorists in psychology in American science. His contributions of epistemology, sensation, perception, pragmatism and imagination are all considered to be cornerstones in American philosophy and psychology. What is rather fascinating about William James was his capacity to conduct research and to describe his findings in such a straightforward and easy to read manner. James was considered a highly gifted writer, even "poetic" by many of his contemporaries and critics and was praised by his unique ability to describe the functions of human consciousness in a very "readable" and captivating format. He described human consciousness as a flowing river, bending, winding and changing in a multitude of ways. He chastised his fellow philosophers for technical writing that proved difficulty to understand. He basically tried to describe Functionalism as a theory that explains how human behaviors have adapted over time and can be influenced by environmental factors, very similar to the work of his colleague Charles Darwin. Ironically, as James pursued his exhaustive work his physical health suffered and he took time off from teaching at Harvard University.

Because of his intellect and keen insight into the mechanisms of environment and behaviors, he often criticized his peers in psychology for making the discipline of psychology unnecessarily complex and abstract. Indeed, he often did not even refer to himself as a "psychologist" as the very thought of the abstraction and deterioration of the discipline increasingly disgusted him especially toward the latter end of his career. As his career developed at Harvard University, William James became increasingly more devoted to the topics of pragmatism as further discussed in *The Meaning of Truth* (1909). The time period and social climate was investigating human behaviors from various areas: Education, science, industry, agriculture and technology were all advancing at unprecedented phases and psychology needed to find a place in these rapid areas of development. Furthermore, rapid changes within the social culture of science itself (i.e., progressivism) allowed scientists and psychologists to pursue a broader range of study in explaining human behaviors. A major goal of the proponents of Functionalistic thought was in creating it as a school different from structuralism's influence during the first half of the nineteenth century. Structuralism was described by Wundt and Titchener as a science of mental structural elements, whereas functionalism was now described as a science of operational thinking.

James' poetic style of teaching made him a very popular professor at Harvard University in 1875, as students would often line up in hallways just to hear his captivating and dynamic lectures addressing human nature and behaviors. He was a gifted writer himself (his brother being Henry James) and was a very creative person who experimented with a variety of interesting and unique activities outside of psychology (he was known to study mental telepathy, psychic powers, and even explored the psychological ramifications of conflict and aggression through war). As an adamant pacifist, he warned his students and colleagues of the psychological influences of conflict, war, and the potential of inhumanity that humans are capable of engaging in. Almost prophetically, he died shortly before the beginning of World War I. James himself was a meticulous scientist who argued that careful observation and detailed analysis were critical elements to the scientific method. William James' characteristics of gathering information in a scientific process set the groundwork for psychology as developing more as a controlled laboratory science.

Additionally, because of James' professional influence in research psychology as a discipline became a more respected science in terms of understanding human behaviors. William James' areas of expertise and specialty in philosophy (i.e., streams of consciousness, sensation and perception) had profound influences in the development of the topics of psychophysics. The late eighteenth century was a period of time where more psychologists were interested in describing the relationship between physical characteristics of the environment (space and area) with human topics addressing sensation and perception.

While teaching at Harvard University in 1875, William James argued that human emotions and feelings cannot be understood nor separated from the environment in which they occur. James tried to describe human behavior as a unique interaction between an individual's constitution and the different types of environmental influences. Different environments could have profound influences on behavior, and therefore the more we understand a person's early history and environment the better we are able in understanding each person's behavior. Environment, DNA and genetics tend to have reciprocal influences upon one another, and only through careful and detailed analysis of both factors can we accurately understand behaviors.

Topics addressing religion and theology shortly after the publication of Darwin's *On the Origin of Species* in 1859 were becoming increasingly criticized as unscientific topics that were simply not worthy of scientific study. William James argued that if religion could somehow provide an answer to the meaning of life, then religion clearly did play a role in the development and adaptation of human behavior. If we consider our purposes in life as a series of ongoing challenges and that we have a moral responsibility to meet these challenges, then James argues that religion plays a logical if not necessary role in the adaptation of behaviors.

Darwin's Influence on Functionalism

William James, as was most of the world, was very much influenced by Charles Darwin's publication *On the Origin of Species* (1859). The relationship between Darwin's (1859) theory of evolution, natural selection and the subsequent development of Functionalism in 1875 cannot be understated. According to William James, consciousness itself provided a useful tool to facilitate survival as it provided organisms with insight to determine how to avoid situations that may have been dangerous or unsafe. Arguably, the "insight" provided by the consciousness also allowed humans to have a broader range of choices to make in critical decision-making strategies, thereby suggesting that human behavior is not essentially reductionistic in nature but rather provides people with free will and gives them the capacity to make rational choices in their lives (Green 2009):

> Taking a purely naturalistic view of the matter, it seems reasonable to suppose that, unless consciousness served some useful purpose, it would not have been superadded to life. Assuming hypothetically that this is so, there results an important problem for psychophysicists to find out, and namely, how consciousness helps an animal. (James, 1875/1980, p. 205 in Green (2009) p. 77)

William James applied some of Darwin's basic concepts of natural selection and variation actually to his own theories in psychology. One important topic that was central to the works of William James was that of instincts. William James argued that humans as a species have several different types of instincts that have helped us survive as a species, including fear, shyness, sucking, crawling, and smiling. Additionally, these instincts can become modified over time when humans develop other complex instincts, including aggression and hunting skills. William James was very much influenced by Darwin's 1859 publication *On the Origin of Species* and it is likely that his detailed description of the several different types of human instincts supports some of Darwin's basic premises of evolutionary theory. The central point that James makes in describing instincts is that they were unlearned mechanisms that evolved to support human survival and given the complexities of human existence during this period of time they were an outcome of the process known as natural selection.

Similarly, consciousness as described by William James clearly existed for a reason—and that was to facilitate how organisms interpreted situations within their environment and how decisions were made. Thus, consciousness too became a subjective construct vulnerable to the same mechanisms as physical forms of evolution within the environment.

Darwin noted in his original work in the Archipelago Islands that an interesting phenomenon occurred between climate conditions, environment and directly observable physical characteristics (*phenotypical traits*) of many species on that particular island. Animals appeared to have developed key physical traits that seemed to facilitate their own survival (i.e., sensory adaptation, physical trait characteristics that would support specific physical environments). Darwin noted that in order to understanding how groups of animals or species were capable of survival, one must first analyze and understand the influences of the environment itself. His famous quotation: "Biology is our destiny" confirms his belief that DNA and genetics have profound influences on our behaviors and the choices that we make, and the factors that influence genetic development occur with environmental changes over extremely long periods of time. Survival was dependent on adapting to climate changes and genetic variations within species or groups of animals.

In essence, then Darwin noted an interesting relationship between characteristics of the environment and specific animal phenotypical traits. He further noted that the different types of environmental characteristics seemed to have shaped or influence the physical appearance of successful animal, fish and insect groups, and that the likelihood of the survival (or extinction) of that species depended on its ability to adapt through genetic variation. Those groups with the greatest chance of survival were also those capable of showing or demonstrating the greatest variation among genes within successive generations. Darwin was also impressed with common links that several different animal groups and species seemed to share, such as internal organs and central nervous systems, a view now referred to as homology. Darwin further argued that all living things needed to adapt and survive a series of threats to their own existence, and that all things must evolve or perish. A key concept that facilitated this process of survival was genetic variation. Because some species were more adapt to change and variation given a variety of different types of climate and environmental conditions, some species were naturally more able to survive and pass on key traits to future generations. This process is now referred to as evolutionary theory and has profound influences on human behaviors and adaptation.

William James was very much intrigued by this radical hypothesis regarding human adaptation and survival, and thus applied these basic theories to his own work in functionalism. If physical environments could influence the physical traits of animal groups, then, according to fundamental functionalist theory, then the physical environment could also shape and influence psychological traits such as cognition, temperament and even intelligence. James argued that thinking, knowledge and cognition were critical elements that helped individuals adapt to their environments and to better understand their life experiences. James basically argued the same point that Darwin was making, but he went further in maintaining that environments not only shape and modify our physical traits, but more importantly our thinking processes and cognitive development. It was a bold step that William James intended to support with his own research relative to introspective observation and emotion.

It is generally agreed that the rapid development of functionalism (late seventeenth century and early eighteenth century) had a profound influence on the demise of Structuralism. Individuals who were now studying psychology were focusing on internal psychological states as being interrelated with other phenomena relative to behavior. Environmental influences were now also considered to be fundamental factors that contributed to the uniqueness of human behaviors. The true ability of understanding psychology, according to the functionalists, was impossible without considering the impact and influences of the environment itself. While the two disciplines fiercely argued over the descriptions and origins in understanding of human behaviors, some psychologists (such as Mary Whiton Caulkins) tried to describe each discipline as complementing each other and not contradicting each other. Unfortunately her efforts were unsuccessful despite working closely with her former professor at Harvard University (William James) and structuralism was essentially a defunct science by the end of the nineteenth century.

When trying to understand human behaviors within their basic elements, Structuralists (such as Wundt and Titchener) argued that learning is a function of internal cognitive processes and structure; the functionalists would argue that learning is more of a dynamic interaction between unique individuals and environments. Differences in learning and adaptation, therefore, are more of a function of the unique interplay between environment and the individual rather than a direct function of internal processing.

The role of Functionalism provided key mechanisms in which psychology advanced in several ways throughout the nineteenth and twentieth centuries. The development of Functionalism as a discipline became possible only through the scientific contributions made by Darwin's theory of evolution and natural selection. Several groups of young American psychologists (i.e., William James, G. Stanley Hall, and Charles Peirce) were gradually shaping the development of a modern new science called Functionalism. This new science had also contributed toward the development of psychology as an applied science, and as a result clinical psychology, developmental psychology and the fields of psychometrics and testing were now beginning to flourish.

RECAPITULATION THEORY: ONTOGENY RECAPITULATES PHYLOGENY

G. Stanley Hall (1844–1924)

William James was primarily known for his contributions to Functionalism and how he developed a branch of psychology that evaluated the role of environmental experiences, learning and adaptation. While he is known for advancing psychology in general throughout the United States, G. Stanley Hall is more credited with innovation in the development of childhood development with psychology as well as introducing an important and related concept to evolutionary theory: Recapitulation theory.

G. Stanley Hall is now considered an eminent psychologist who was more concerned with understanding the relationship between environmental and social context and learning. His lifetime contributions to psychology were enormous during the late eighteenth and early nineteenth centuries, as he was elected as the first president (and founder) of the American Psychological Association and was actually the first student to receive the PhD in psychology. Hall was also the first president of Clark University and helped establish this university as one of the premier institutions of higher education in the eighteenth century. Hall's real accomplishments, however, were more about his driving force regarding evolutionary theory and recapitulation theory.

Hall was also noted to support the evolutionary theory as proposed by Darwin and he used the brain and nervous systems as key components in support of the evolutionary theory. Perhaps one of Hall's most significant contributions was his "recapitulation theory." The recapitulation theory (sometimes also described as "ontogeny recapitulates phylogeny") describes a broad range of animals and organisms that develop in similar sequence and stages, dating back to early evolutionary history. Early in the prenatal (i.e., germinal and embryonic) phase of development, humans reflect a more primitive phase of development that seems to match other more primitive species. For example, Ernst Hackel identifies several observable phenotypical traits (i.e., pharyngeal grooves) during embryonic maturation that seems to reflect more of a fish-like appearance (see Figure 3.1 below) that is representative of early human evolutionary processes (Richards 2008).

Recapitulation theory as described by Hall also argued that young (preadolescent) children should not be confined to strict or confining rules during development, but rather they should be afforded ample opportunities of physical growth, activity and exercise within the environment. Additionally, Hall felt that young children should be provided with environments that encouraged physical development that would simulate physical challenges that humans may have been typically exposed to during our early (primate) evolutionary phase of development. During this period of time (late eighteenth century to the early nineteenth century), the prevailing view regarding children's behaviors and learning was that cognitive development should be structured, allowing little creativity and individual development. Hall disagreed and argued that children learn best when they are provided with flexibility and allowed to progress from various stages individually. Soon after Charles Darwin published *On the Origin of Species* in 1859,

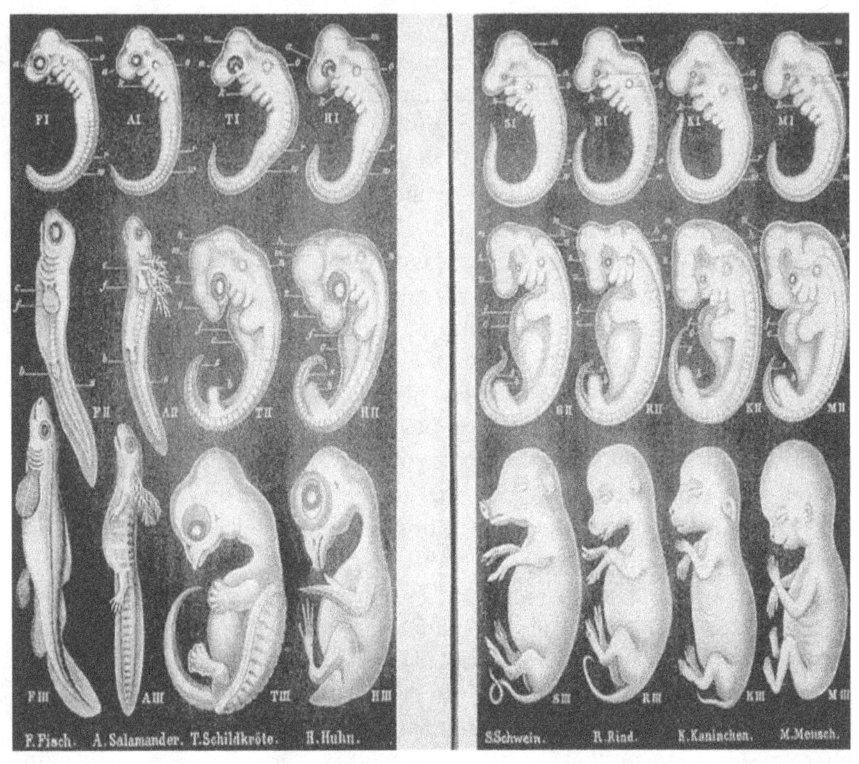

Figure 3.1 Ontogeny Recapitulates Phylogeny. *Source*: 1874 illustration from Anthropogenie showing "very early", "somewhat later" and "still later" stages of embryos of fish (F), salamander (A), turtle (T), chick (H), pig (S), cow (R), rabbit (K), and human (M).

Hall was immensely interested in this new idea regarding human behaviors and approached it from a psychological perspective.

The recapitulation theory as prescribed by Hall in 1904 (*Adolescence: Its Psychology and Its Relations to Physiology, Anthropology, Sociology, Sex, Crime, Religion and Education*) argued that a correlation between stages of *individual* human development exists with how humans have evolved *as a species* over the millennia. These patterns of behavior have repeated themselves over thousands of years, beginning with infancy where behaviors may be categorized more as primarily egocentric and amoral with gradual stages of development where humans learn to develop the qualities of sharing, cooperation and (in some instances) altruistic behaviors as being essential for healthy community development. Humans have evolved over long periods of time from a selfish, "savage beast" to a species capable of making sacrifices to improve the overall human condition. As a species, then, Hall argued in his recapitulation theory that this pattern of growth from self-centeredness to selfless behaviors, including the capacity for love and wisdom, repeats itself with every generation. Hall compared the early, crude and illogical behaviors of the young child to the early evolutionary history of humans as a species. Over time humans gradually acquired the use of social skills (i.e., cooperation) as a means of improving survival and community development. By late childhood Hall argued children need to be exposed to formal education, moral reasoning and essentially taught the skills required in community living (i.e., sharing and exchanging information, cooperative relationships, etc.).

In the beginning of his academic career in 1863, Hall first enrolled in Union Theological Seminary in New York as a means to continue his work in teaching and education. The choice however was ill-fated as his views addressing the biological and evolutionary role of humans increasingly conflicted with the more traditional religious theories offered by Union Theological Seminary. Although still intrigued in evolutionary theory, Hall needed to restrain his enthusiasm or risk expulsion from his school. He continued to tutor students in literature and he also taught education courses relative to learning theory. Hall was the first psychologist to determine the educational principles of learning theory in psychology and argued that environments in the classroom can stimulate learning among younger students. Hall's work while at Union Theological Seminary was highly instrumental in describing how children learn and what educators can do to stimulate learning at the primary grade level.

Hall was also influenced by Wilhelm Wundt's important contribution: "Physiological Psychology" which was published in 1874. The book was a tremendous success and explained the physiological relationships between psychology and human behaviors. This fundamental concept addressing behaviors, physiology and adaptation also helped to connect functionalism with the increasing popularity of evolutionary theory. Because of Hall's tremendous success and popularity as a tutor he was offered a part-time teaching position at Harvard University in English. His teaching experiences at Harvard University became a central influence in his work in evolutionary psychology, as he was able to correspond and work with other famous psychologists such as William James. Hall enrolled as a graduate student at Harvard University and graduated in 1878 with the first doctoral degree in the United States.

As Hall's work and teaching became more popular, he was offered a full-time teaching position at Johns Hopkins University in Baltimore, MD. Hall's dynamic lecture style and interesting topics addressing the theories of learning, education, and the very new theories of evolution made his classrooms always very full and popular among students.

Current evolutionary theorists would support one of Darwin's most basic assumptions in his original work was that any adaptive behavior that supported survival skills tended to be retained throughout generations and those behaviors that were not considered adaptive eventually became obsolete. G. Stanley Hall noted in his recapitulation theory that all children, regardless of cultural or ethnic backgrounds, do share very common similarities in their gradual development of behaviors. These behaviors become "recapitulated" over time with successive generations and show striking self-centered and egocentric characteristics over time. Hall argued that a primary aim of responsible parenting, therefore, was in teaching children concepts relative to sharing, cooperative behaviors and perhaps most importantly the capacity to delay gratification over time.

Hall characterized these early adaptive behaviors as "savage like" and that they were designed to help infants survive by maintaining a highly egoistic and self-centered approach to the world. The infant does not recognize the needs or even existence of others (at least in the first two or three years of life) as it is virtually impossible for them to do so. This is the epitome of egocentric behaviors—an inability to recognize the needs of others in the world around them. Over time, however, Hall notes that even the most self-centered children can be taught selfless and caring behaviors that exist and recognize the needs of others. These behaviors that are gradually learned by children are developed gradually, and slowly, and are exhibited cross-culturally.

Hall's innovative recapitulation theory seems to support what many early developmental theorists have argued for many years, in that many adaptive behaviors among children are manifested in sequences or stages, depending on levels of stimulation and interaction within the environment. For example, crawling, standing and walking are typically demonstrated within 6–18 months. Cognitive growth and development also seem to develop similarly among children from diverse backgrounds. Terms introduced by Jean Piaget, including object permanence and conservation do occur cross-culturally and typically develop during early childhood (i.e., 3–8 years). These cognitive skills not only help children to better understand and master their immediate surroundings, but also help them to establish meaningful relationships with others in a social environment. The common physical and cognitive forms of development have repeated or "recapitulated" themselves for generations as a means of improving overall adaptation and development. Civilization, education and helping children to delay their *perceived* needs of immediate gratification via socialization are essential qualities in establishing the foundation of a civilized society and are common among all worldwide cultures.

The works by G. Stanley Hall became important theories of both development and psychology and served as a primary influence to theorists such as Jean Piaget, John Dewey and William James who later developed the now famous theory of Functionalism. Hall's recapitulation theory was highly innovative during this period of time as it shaped how theorists described human behaviors as occurring in sequences and stages defined by human evolutionary behaviors. Perhaps one of the most important influences that functionalism had in psychology was the development of a more practical and applied approach to the general public. Psychology was now being taught more often in American universities, but in order for it to become one of the leading disciplines in mainstream American cultures, people needed to see how the basic concepts of psychology could be applied to daily human interaction. The lack of practical application to problems in everyday life was a major flaw in the structuralism school of thought. Evolutionary psychology and the practical applications of functionalism offered in the nineteenth century helped to create psychology as a modern science.

The influence of the development of evolutionary theory and psychology was rapidly growing in the United States shortly after Darwin published *On the Origin of Species* in 1859. Functionalism also played a vital role in the development of evolutionary theory as both schools of thought explained behaviors from very practical concepts. The relationship between Structuralism and Functionalism could not have been more different. For example, functionalism emphasized how behaviors are learned and shaped through environmental experiences, whereas structuralism was more concerned about introspection and units of consciousness. Additionally, functionalism was rapidly becoming a science that emphasized methodology, measurement and empirical evaluations, something that simply did not occur with the philosophy of structuralism. An additional area of controversy was that of animal intelligence. Darwin defended this concept vigorously by noting that it would be egocentric and foolish to consider humans as the only species capable of intelligence and reason. Animals have been provided with a series of experiments and problems that can at least suggest that some form of adaptation and intelligence facilitated their existence and survival. In his publication: *The Expression of the Emotions in Man and Animals* (1872), Darwin makes several arguments about how human facial expressions (i.e., smiling, frowning and sneering) historically have played critical roles in human survival (even today most theorists argue that basic emotions and facial gestures are in fact universal).

INFLUENCES OF EVOLUTIONARY THEORY IN MODERN PSYCHOLOGY

Charles Darwin (1809–1882)

Perhaps one of the greatest challenges facing Charles Darwin regarding his publication: *On the Origin of Species* in 1859 was not so much the scientific community, but the reactionary forces from earlier schools of thought (i.e., Structuralism) as well as more conservative religious influences both in Europe and the United States. A strong supporter of Darwinian Evolutionary theory and personal friend was Thomas Henry Huxley (grandfather of *Brave New World* author Aldous Huxley). Huxley was a well-regarded biologist in Europe and helped to organize the now famous debate sponsored by the British Association of the Advancement of Science and held at Oxford University in 1860. The conference had helped to establish Darwin's work not so much as an attack on fundamental religion but actually as advancement in scientific thinking and understanding modern psychological and human behavior. So great was the pressure not to publish any work that challenged the conservative norm of institutionalized religions, Darwin considered abandoning his work out of fear of a religious community backlash. As it was, he confided to a friend that he felt as guilty as "confessing to a murder" in his discussions of the evolutionary theory (Desmond 1997).

Darwin's aim was not so much to suppress or contradict the existence of God or religion in so much as he attempted to portray a rational, *ipso facto* explanation of the origin of the human species. While not a self-proclaimed atheist, Darwin merely attempted to provide a scientific interpretation of the origins of humans without reference to mystical entities or spirituality. His best and most successful effort in identifying the evolution of humans may have actually been relative to his field work addressing the relationships between animal behaviors and human behaviors in his publication: *The Expression of the Emotions in Man and Animals* (1872).

In this text, Darwin argues that a primary motivator of human behaviors is essentially emotions that humans have inherited from lower species. The emotional expressions that are now common to humans (i.e., smiling as an indicator of a positive emotions or frowning as an indicator of a more negative emotion) have recently been determined to be inherent and cross-cultural (Frank, Ekman, and Friesen 1993).

Darwin's work, while a huge success in Europe in the late eighteenth century, took significantly longer to catch on in the United States. Darwin attempted to describe the function and purpose of human emotions as adaptive in human survival, and that many human facial expressions are in fact universal (Ekman and Friesen 1986; Ekman, Friesen, and Ellsworth 1972). A grimace or frown, for example, has been shown to reflect anger throughout any global culture, while a smile has also been shown to universally represent happiness or pleasure.

As previously stated, Darwin's work had tremendous influence on a number of contemporary theorists. His theories addressing the emotions of human behaviors were different than some of his contemporaries, such as William James. Psychologists were historically devoted to understanding how and why people experienced "feelings" or emotions, and how our bodies responded to events that may have triggered fear, arousal, love, etc.

Darwin was not the first theorist to discuss how different types of species adapted to the harsh demands of the environment. Jean Baptiste Lamarch (1744–1829) argued that some of the physical traits possessed by some organisms are due to physiological changes within their nervous system due to conflict. As a result of this conflict, fluids emitted during battle caused specific organs to change in size and texture. Darwin's work in 1859 really described how these biological changes took place over extended or longer periods of time, a process that he referred to as natural selection.

Darwin's work centered on the theory that all of life remains in a continuous battle or struggle for survival. Humans were not the only living creatures or animals that engaged in a battle for survival; all living things, including the plants, trees, and fauna in general not only battled for their own existence, but were quite adept in producing toxins that would prove highly distasteful to would-be predators. In his short but brilliant summary of traditional Darwinian evolutionary theory, David Buss (2009) summarizes Darwin's work into three primary areas of adaptation: a) Struggles with the natural conditions of life; b) Struggles with other species; and c) Struggles with members of our own species (see Figure 3.2 below).

Human Existence was a Constant Battle for Survival

Darwin knew early on in his exploration of the Archipelago Islands that the plants and animals competed fiercely for survival. Trees loomed vertically skyward for several hundred feet in fierce competition for sunlight; plants and trees adapted by growing spikes and thorns in circumference of their trunks as a means of warding off insects and birds. Key resources that are vital for survival were available only to those individual members of a particular species that showed cunning, skill and resourcefulness to acquire these resources if they lacked the brute strength to overpower competitors. The biological struggles with the natural conditions of life proved to be a pivotal point in human evolution that defined specific characteristics of certain environments. Researchers, for example, have shown universal preferences for certain forms

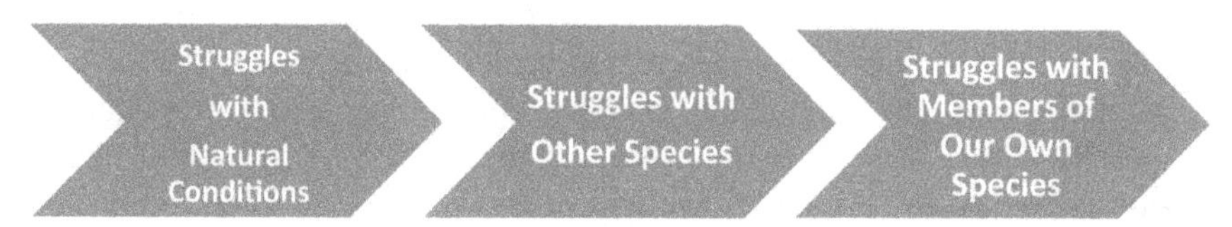

Figure 3.2 The Biological Struggles of Evolution.

or types of environmental terrains that provide most conducive to supporting life. This would include an abundance of fresh or potable water, green shrubs to provide cover from predators as well as a form of food to consume, and a universal preference for certain types of fruits, including berries and green environments (Orians and Heerwagen 1992).

Buss (2015) has also identified three conditions necessary for natural selection to occur: Genetic variation, inheritance factors and overall reproductive success (p. 5). Without genetic variation, humans as a species would not have survived. For example, warmer climates require darker skin pigmentation as a safeguard against skin cancer, and those organisms with genetic traits promoting these phenotypical traits would have survived over the course of time and passed on those traits to future generations. A persistent and lingering problem for Darwin's theory of natural selection was the peacock. A peacock has brilliant plumage of feathers that clearly would alert any predator. The idea of an organism deliberately exposing himself or herself to a potential predator clearly contradicts the theory of natural selection, as this tendency would reduce (not increase) reproductive fitness. Sometime after his theory of natural selection, Darwin introduced the concept of sexual selection. Sexual selection refers to a form of adaptation that is designed to attract potential mates for reproduction. So, while it is true that an attractive and colorful assortment of feathers is risky in terms of attracting potential predators, the colors could also attract several mates that would significantly increase future linage of offspring and reproductive success.

Interestingly, humans have also demonstrated different types of universal behaviors that have evolved through natural selection. Environments that posed unique natural risks, such as cliffs, large bodies of water, or areas prone to floods and earthquakes were typically avoided and considered hazardous to healthful living conditions. Human adaptation and survival was not limited to the environment. Humans quickly discovered in their hunting efforts that if they wandered too far from points of safety and refuge that they too could become prey to other predators. Perhaps one of the most menacing threats came from neither the environment nor other species, but ironically from our own species. Warfare, territorial disputes and interpersonal conflict tragically are a common event among humans. Early humans engaged in conflict over a variety of resources, but it appears that one of the most volatile resources was that of reproduction. Early human males (especially the Alpha male) typically engaged in a variety of different types of battles of females and the ability to sire offspring. Successful males in combat were thus able to impregnate more females and thus increase reproductive fitness. Research by (Buss 2009; Zerjal, Xue, Bertorelle, Wells, Bao, and Zhu 2003) has shown that blood samples taken near the Mongolian empire represent an 8% chromosome match of successful Mongolian warriors. What this means simply is that the warriors who were most successful in combat (such as the Mongolian warriors under Genghis Khan) were first able to terminate male competitors of a given area, then abduct and impregnate as many fertile women as possible. Thus continues the spoils of war.

Chapter 4

Three Primary Behavioral Influences in the Development of Evolutionary Psychology

Thorndike, Watson & Pavlov

An interesting theme within the development of modern psychology with evolutionary psychology is the role that the characteristics of the environment have in the development of both phenotypical traits and the process of learning. The Behaviorists (beginning with John Watson) argued that the events occurring within the environment shape what we learn as well as our personalities. Modern evolutionary p-psychologists have argued that the conditions of the physical environment have shaped not only who we are in terms of physical appearance (phenotype), but interestingly also how we socialize and interact with one another.

During the early developments of the nineteenth century, the science of Behaviorism was rapidly becoming a strong political, economic and educational reaction against the contemporary views of Structuralism and Functionalism. Behaviorism brought a refreshing new perspective and clarity in understanding how people behaved within different environments and had strong influence among contemporary intellectual thinkers and educators. The discipline of Behaviorism was more of a cultural revolution that provided a practical explanation of people's behaviors rather than simply an academic discipline describing individual "thoughts." Private enterprise, corporations, schools and governmental agencies now were requiring psychologists to be able to empirically validate the marketability of products by way of psychological innovation. Psychologists using the Behavioral approach were now being incorporated into the public and private sector to share their skills and help business and enterprise expand in the community.

The new school of Behaviorism brought changes in how psychology was to be applied as a science—testing, psychometrics, counseling, clinical psychology and industry were new areas employing the use of the Behavioral principles of psychology. One of the pioneers in psychology that helped advance scientific thinking through the development of the basic principles of learning was Edward Lee Thorndike.

EDWARD LEE THORNDIKE (1874–1949)

Considered by many to be one of the most important and an influential thinker in the development of comparative psychology and Behaviorism, Edward Lee Thorndike was a leading researcher who attempted to explain learning through experimentation with animals under controlled testing conditions. Thorndike worked closely with William James while at Harvard University where he earned his Master's Thesis addressing the topics of animal behavior and animal learning. It was here where Thorndike conceptualized the idea of attributing the process of animal learning with humans. Thorndike is credited with being the founder of comparative psychology and development of formalized animal research procedures.

Much of his work has contributed to the development of learning theory and educational psychology. Thorndike's background combined educational learning with psychological theory and created a fundamental approach to understanding the mechanics and characteristics that best influence the dynamic process of learning. Thorndike continued his area of study with animals and completed his PhD ("*Animal Intelligence: An Experimental Study of the Associative Processes in Animals*") at Columbia University under the supervision of James McKeen Cattell. Thorndike's work in psychological represented a break from the more traditional Functionalistic school of thought and introduced the tenet's of Behaviorism.

Thondike initially wanted to complete his research in learning and education with children, but he was unable to acquire human subjects while completing his Master's thesis at Harvard University in 1897. He was provided with more academic resources and accommodations by James Cattell at Teacher's College (Columbia University) and completed his dissertation in 1898. Cattell's own research in measurement and psychometrics had an important influence in the development of Thorndike's own research addressing the quantification and measurement of animal learning and instrumental conditioning.

Law of Connectionism, Law of Effect and Law of Exercise

Perhaps Thorndike's greatest accomplishments in the discipline of psychology was his ability to predict and define and describe the conditions in which learning occurs most efficiently, and identify how animals and humans learn in very similar ways. Considered by many to be his greatest contribution in psychology is the theory of connectionism which was determined through a series of experiments involving cats in the development of the puzzle box.

Connectionism was a theory describing the bond or "connection" between a stimulus and a response. This bond or connection of learning is strengthened as one's behaviors are reinforced via positive or favorable consequences. When a cat (in a motivated state to escape) is placed in a box with a lever that needs to be activated to escape, once the lever is activated the behavior is more likely to be repeated within a shorter period of time. The discovery of the exact responses necessary to activate the cat to its freedom and access to food was neither sudden nor dramatic. The types of learning based on Thorndike's observation of the cat were progressive and incremental. This process of "trial and error" learning allowed the organism to discover what types of behaviors were necessary to activate the lever which in turn released the spring door for the cat. The positive consequences (i.e., food) which were a result of a specific action (pressing a release lever) resulted in a very positive state of consequences.

The law of effect argues that any behavior that results in a satisfying reaction to the organism is most likely to repeat itself in similar situations. Thus, a child who is rewarded for studying and improving his or her grade point average may be included to increase studying within a variety of different types of situations and environments as a general means of improving his or her academic work. Human behaviors were most adaptive if they helped individuals to not only survive, but to flourish in harsh environmental conditions. If a behavior is particularly productive in that it produces a necessary resource for survival, the law of effect argues that it will be repeated again in the future within similar types of situations. Similarly, the law of exercise argues that as the number of repetitions increases relative to behavior, so does the overall strength of the relationship between the S-R (stimulus—response) connections.

Thus, the process of learning was strengthened between the behavioral stimulus of the cat and the subsequent consequence (access to food). This process of learning was both measurable and predictable based on the observations of Thorndike which he described as a "learning curve." The demonstration of the learning curve also helped Thorndike to disprove that learning was not a sudden awareness of previously unrecognized relationships (as the Gestaltists would later argue), but rather discovery and retention of knowledge is based on a series of activities and experiences (later referred to as "instrumental learning") commonly described as the process of trial and error.

The overall impact and ramifications of Thorndike's contributions addressing the theories of learning through Connectionism cannot be overemphasized. Thorndike was the first (even before Pavlov) to identify the circumstances and conditions that facilitated learning and adaptation under new environments. Thorndike argued that the effects or consequences of particular behaviors were not necessarily the only factors needed for learning. The "bond" or the connection of learning was strengthened and became associated with particular characteristics of the physical environment. If the consequences of a particular type of behavior (i.e., a cat pressing a specific lever) are rewarded (i.e., escaping to allow access to food), then that behavior is more likely to repeat itself in the future. Similarly, those behaviors that are associated with negative consequences (i.e., punishment), then those behaviors will be reduced. This theoretical framework would soon become the foundation of Behaviorism and more specifically influenced much of B. F. Skinner's work in operant conditioning.

Thorndike was also able to disprove the observational theory of learning through his experimentation and work with cats in the puzzle box. Thorndike initially hypothesized that, like humans, primates and mammals could learn simply through observing other organisms engaged in a variety of different types of behaviors. Thorndike discovered that the groups of cats that were exposed to other animals pressing a lever to escape the puzzle box were no more efficient in escaping from the puzzle box than cats who were not exposed to the group of cats escaping from the puzzle box. Thorndike even placed the legs and paws of the cats on the lever of the puzzle box in an effort to "teach" them how to escape. These efforts were also unsuccessful in improving the rate of learning for escape. Thus, Thorndike concluded that the process of learning (at least for cats) is dependent on an environment that allowed the organism to learn by way of trial and error and experience.

Once the cat learned specifically what actions were necessary in activating the escape lever, these actions provided a "connection" of learning that produced a "satisfying state of affairs" between the action (pressing release lever) and the consequence of the behavior (exiting the puzzle box and gaining access to food). Thus, in summary, some of the most fundamental and basic laws of learning were discovered by Thorndike in the research involving the puzzle box:

- All learning is incremental or "step-by-step" and not sudden learning often referred to as insight learning;
- All animals learn the same way (i.e., through the law of effect) and what they learn becomes automatic; and
- If a particular behavior produces a positive outcome, then the bond or the "connection" between the stimulus and the subsequent response is gradually strengthened. This phenomenon is what Thorndike referred to as the "Law of Effect."

Thorndike's research essentially served as a catalyst in the future development of Behaviorism, and more generally the theories of learning relative to psychology. Even Pavlov commented on the breadth and development of the work that Thorndike completed and how his work was the primary influence in the development of Behaviorism. More importantly, however, was the key component of the relationship to learning based on Thorndike's research with animals. Thorndike's research involving his now famous puzzle box serves to illustrate the dynamic relationship between environment, intellect and adaptation. Those key behaviors that were useful in helping animals to survive were essentially "stamped in" due to their frequent and important use, whereas those behaviors that were less useful became weakened or "stamped out" and eventually extinct.

Those animals that were more capable of learning faster and under more diverse conditions of the physical environment were clearly able to survive over those other animals showing less adaptive survival strategies. A cat's paw activating a mechanism for release and access to food was clearly an instrumental tool useful for catching prey, climbing and escaping from predators. The physical characteristics of the environments shaped not only the gradual phenotypical characteristics of the organism but also the intellectual and cognitive resiliency of the organism in its efforts of survival. Thorndike's puzzle box research not only served to facilitate the future development of Behaviorism in modern Psychology, but his research also provided the strongest support yet for the case in the development of evolutionary theory as proposed by Charles Darwin.

Thorndike's research with cats showed that learning for most animals does not occur through insight but rather over time and is a gradual process. Related contributions to the discipline of psychology are straightforward and simplistic but serve to explain the process of learning in a fundamental way. The "law of use" was offered by Thorndike describing that the more an association is used the stronger the connection of the association becomes. The "law of effect" similarly argues that if a specific behavior is followed by a "satisfying state of affairs" then the bond of the relationship is actually strengthened, and conversely if a behavior is followed by a "dissatisfying state of affairs" then the bond or connection of learning is weakened. Thorndike argued that all animals (including humans) learn in similar ways, and that the more we understand the process of learning for people the more able we are to improve and change behaviors to a "satisfying state of affairs":

The time period of 1914 brought with it a strong reaction against traditional European thinking. Western thought was now replacing more traditional views (i.e., introspective) of human behavior. The Austro-Hungarian invasion of Serbia and the Russian invasion of Germany culminated in a world war that came to be known as World War I. The United States, under the political leadership of Woodrow Wilson, had large investments in the European trade market, notably England and France. There was no current desire to become involved in a "European war" where the United States was profiting by the development of war goods to other nations.

President Wilson's efforts to make the "world safe for democracy" with the sinking of the Lusitania and the Russian Bolshevik Revolution of 1917 all contributed to the United States reluctantly becoming embroiled in World War I. These rapid political and economic changes soon resulted in change in how Americans viewed causal influences of human behavior. A growing academic interest spawned important questions that explored causal factors that were associated with human aggression and conflict, and conversely, human prosocial and altruistic behaviors. Could environmental forces induce individuals to conform to crimes against humanity? Or, as Freud indicated in his publication of "Civilization and its Discontent" could aggressive and anti-social forces be instinctual forces that define our personality?

The current intellectual climate of the early nineteenth century and higher education in general now identified environmental factors (i.e., rewarding experiences, antecedent conditions, reinforcements, etc.) as key components in our ability to predict behaviors. A machine-like and mechanistic approach to understanding human behavior was now considered to be central to the scientific approach. The intellectual zeitgeist of the early nineteenth century was more about discarding innate concepts that attempted to delineate human behavior and more about predicting human behaviors with scientific accuracy based on environmental influences.

Rather than using methods such as "introspection" and "molecularism" in describing human behaviors, the Behaviorists argued in the futility of describing something that cannot be physically or empirically validated or measured. The Behaviorist approach was an effort to structure psychology more as a scientific and measurable science rather than an imprecise discipline that relied on self-reports in understanding behaviors.

JOHN B. WATSON (1878–1958)

While Edward Lee Thorndike was considered a pioneer in the animal research and learning, John Watson's research addressing conditioning principles created a foundation for the development of Behaviorism. Indeed, much of Watson's work has been described as the "manifesto" of the development of theories addressing Behaviorism ("*Psychology as the Behaviorist Views It*"). The beginning of the Behaviorist movement coincided with the demise of Structuralism and introspective thinking (and many would later argue even helped cause the demise of the discipline of Structuralism). Part of the reason of the rapid popularity of Behaviorism was its sense of control and autonomy over human behavior.

The goal of all Behaviorists is simply the mastery, prediction and control of human behaviors. Most Behaviorists would argue that positive and negative human behaviors come from our own control—entirely an internal locus of control. Watson described Behaviorism as primarily an objective science that only allowed for the measurement of overt behaviors. Thoughts, emotions, attributes and so-called traits and dispositions were rejected as subjective constructs that merely confuse and cloud accuracy within the scientific world.

In order to truly understand how John Watson developed Behaviorism, we need to review his background and history as a small child. Watson was born in a small city in South Carolina. He was reared in a highly unpredictable and contradictory home environment, where his mother, Emma Watson, was a devout Baptist who lived a frugal and stoic life. His mother did not allow nor approve of dancing, alcohol, or smoking anywhere in their home. However, Watson's father (Pickens Butler Watson) was the antithesis of a strict religious culture. He was an unemployed alcoholic father who was also known as a womanizer. His father left the family when John B. Watson was only 13 years old. The experience of poverty left an indelible mark on Watson, who vowed never to be poor and work tirelessly to achieve success in psychology.

Perhaps one reason why Watson emphasized control of environmental conditions as being so important in understanding human behaviors was that he simply never had any of it while being reared as a child. Watson attended Furman University as an undergraduate student, but had difficulty in establishing social relationships and finding a discipline that captured his interest. While at Furman University, Watson did gain the confidence of one of his faculty, Dr. Gordon Moore. Watson applied to the University of Chicago and graduated in 1903. His area of research specialty, understanding the process of the development of the nervous systems in rats, provided the subject and impetus for his dissertation: "*Animal Education: An Experimental Study on the Psychical Development of the White Rat, Correlated with the Growth of its Nervous System*" has been suggested as an important precursor to his later views in the development of Behaviorism as well as his rejection of the existence of instincts. In this dissertation, Watson describes all learning and muscular behaviors as a function of the myelinization in the brain. Rats learn specific sequences in mazes through kinesthetic development. Human behaviors that are often perceived as instinctual are actually learned behaviors through the myelinization process.

Watson's approach and philosophy was certainly not new, as even the early philosophers such as Rene Descartes argued that before any true "science" can develop, first an understanding of the environment and human experiences are essential. John Watson studied under the direction and supervision of James Rowland Angell. Watson's views addressing human behaviors changed the direction and focus of psychology, as human behaviors were considered to be a direct result and consequence of actions within the environment. Watson's ability to radically change the course and direction of psychology was a result of several unique circumstances culminating together as a "perfect storm." The first factor was the overall culture or "zeitgeist" of the era. Academia was quickly changing in the sense that people from different educational and economic backgrounds wanted a more scientific and accurate explanation surrounding human behaviors in general. The "creationist" theory explaining the origins of humans was quickly losing favor among Americans from various cultural and ethnic backgrounds, especially since the publication of Darwin's 1859 *On the Origin of Species*. The general public was looking for a science that would help them to regain mastery and control of their own behaviors.

Behaviorism provided a unique explanation for people to better understand their own behaviors through analysis of environmental conditions. Additionally, Behaviorism proposed the refreshing view that humans were in fact capable of changing their negative or self-destructive behaviors to positive and productive behaviors. Traditional and common problems such as depression, gambling, alcoholism were now being viewed as changeable conditions pending analysis of the environment which purportedly caused these problems.

A second factor that led to the rapid development of Behaviorism was the economy itself. By 1919, World War I had quickly transformed psychology from a secluded academic anomaly to a more relevant and practical science. Psychologists were now more active in exploring ways to measure human performance, skills and aptitudes as a way of succeeding in the war. Subsequently, testing and measurement also became important tools of the psychologist. Finally, a third contributing factor that helped Watson change the scientific world was simply timing. Soon after his arrival to Johns Hopkins University, his predecessor James Baldwin (Chair of Psychology) was quickly removed due to a scandal involving a house of prostitution. The position of Chair of Psychology was awarded to Watson at the very young age of 31. His drive and determination to succeed (he was the youngest person at that time in the history of the University of Chicago to receive his PhD at the age of 25) helped promote him to world fame in a few short years.

There was another educational phenomenon that helped accelerate the rapid popularity of the discipline of Behaviorism. The Mechanistic approach strongly influenced the development of Behaviorism, as human behaviors were considered to be more of a stimulus—response function. The Behaviorists further argued that it was virtually impossible to measure any mental or cognitive event, and that in order to recreate psychology as a "pure" science, all data must be empirically and objectively validated. Terms such as "consciousness" and "introspection" were clearly not a part of the Behaviorist's vocabulary due to their subjective and arbitrary nature.

Historians in psychology have also identified several contributing factors that ultimately led to the formation and development of Behaviorism as a science in psychology: a) The increased role of animal experimentation; b) The reduction of the role of introspection and subjective reports of experiencing sensations; and c) The increased role of Functional Psychology. More research was now being conducted on rats and the rate of learning in mazes, and individual psychologists were now being associated with specific types of animal experimentation.

Much of the work of other psychologists (i.e., Edward Lee Thorndike's earlier work addressing precise measurements) primarily served as an incentive for future Behaviorists (such as Watson) to create a new science of psychology that measured behaviors primarily as a function of environment consequences. Thorndike's work involving cats helped to explain how certain types of consequences of events can shape human behaviors. In many ways, the public was ready for a new interpretation and explanation of behaviors that was based on the groundbreaking work of Charles Darwin. Psychology answered that call by developing a more applied approach, showing the world that psychology has a place outside of the laboratory in helping people to modify and shape their behaviors in a more productive manner.

Since Darwin's publication of the *On the Origin of Species* (1859), more research was being conducted involving animals, intelligence and humans and this research supported the inclusive view of intelligence as a continuum among all animals rather than an exclusive entity. Thorndike argued in his law of effect that the consequences of an action are the greatest determinant whether or not that behavior is likely to be repeated in the future. Many of the leading Behaviorists, such as John Watson, Ivan Pavlov, E. L. Thorndike and B. F. Skinner argued that the primary factor necessary in predicting human behavior is not the genetic or biological make up of an individual but simply an understanding of the antecedent conditions and consequences of specific behaviors. Negative and antisocial behaviors such as crime and violence repeat themselves not because of a "genetic predisposition" towards violence, but rather because of a system of rewards.

Many societies and cultures unwittingly reinforce the very behaviors that they claim they wish to extinguish. The primary focus in understanding human behavior was in addressing the environment and the effects or consequences of individual behavior. Darwin argued that through the process of natural selection, those behaviors that were most adaptive (i.e., those most likely to facilitate survival and well-being) would most likely be passed on to future generations. Environmental traits and characteristics "select" physical qualities that facilitate survival. According to the Behaviorists, consequences of behaviors within specific environments are the greatest determinants of future behaviors. The Behaviorists and evolutionary psychologists had much in common as physical traits and qualities of different environments were considered to be the most important factors in selecting traits that facilitated survival among humans. Watson's approach to understanding human behavior was in removing any question of the causal factors that may have actually influenced behavior.

In other words, only physical and observable behaviors were considered to be appropriate areas of study in psychology, similar to Auguste Comte's views of science in Positivism. Watson argued that consciousness was a fallacy, a contrived idea that could not be objectively studied. For these reasons he is considered to be a radical Behaviorist, and he attempted to reconstruct all of psychology as a science that was measured purely through overt behaviors. Even the concept of "instincts" (views that are commonly accepted today in biology, ethology and evolutionary psychology) was rejected on the grounds that they were simply emotions that were reinforced or rewarded. A child's crying (typically thought of as instinctual) becomes continuous or exacerbated by parents providing excessive attention (i.e., rewards) to the child. Watson's views on child-rearing and child development were considerably stoic and unpopular (he supported a highly authoritarian approach to child-rearing) and he supported a very emotionally detached relationship with children. He published a text offering advice to parents with child-rearing problems (Psychological Care of Infant and Child) in 1928, but he later admitted that child-rearing was not his specialty and that he "did not know enough about the topic."

Instincts, then, according to Watson were merely behaviors that were rewarded and so thoroughly ingrained in behavior that people confuse them as being "innate" or inherent. For example, those behaviors most likely to help survival of the organism are thus rewarded with greater access to key resources (i.e., food, reproduction) and therefore more likely to reoccur in future generations. Environmental influences changed the way humans thought and behaved. Environmental factors also influenced not only the phenotypical characteristics among humans but also the intellectual and cognitive skills that clearly facilitated survival. The relationship between evolutionary psychology and Behaviorism lies in the environment itself; controlling environmental conditions will ultimately control behaviors demonstrated by the organism. Controlling antecedent conditions and their consequences will also determine whether or not behaviors are continued or discontinued. The reason why Watson's views of Behaviorism held tremendous appeal to the general public was that it provided a clear methodology in understanding human behavior—both a scientific and measurable process that provide a method for individuals to shape and control their behaviors and therefore take control of their destiny.

Much has changed within the general discipline of psychology and Watson's radical views regarding Behaviorism. Watson was trying to remove the subjective "impurities" of human emotions and affect in an effort to more accurately understand human behavior. Two important changes in Behaviorism since Watson's (1913) contributions address how experimentation occurs under controlled laboratory conditions and how these proposed changes in behaviors are measured (Kimble 1994). The construct of "truth" holds different meanings for different professions and disciplines. In theology and religion, "truth" is determined through interpretations of scripture and the word of God; in philosophy, "truth" is measured in how we define meaning within our own lives, and for the Behaviorist, "truth" simply refers to an accurate understanding of causality within the physical world as well as an objective assessment of the impact one has had in the world in which he or she lives.

JOHN WATSON, ROSALIE RAYNOR, AND THE LITTLE ALBERT EXPERIMENT (1920)

Arguably the most controversial of all experiments conducted in psychology, the now infamous "Little Albert" study of 1920 sought to condition emotions in humans, thereby further supporting the basic tenets and principles of Behaviorism. The year was 1920; Watson was trying to create a name and reputation for himself through his research in Behaviorism. His world was changing rapidly, and he soon became overwhelmed with the power and prestige that he was rapidly gaining through his academic and professional position at Johns Hopkins University. As with so many other leaders with tremendous future potential, Watson became oblivious to his indiscretions with students (he carried on an affair with his then graduate student and researcher Rosalie Raynor) as well as his indifference to the welfare of the participants in which he carried out his experiments.

"Albert" (actually a pseudonym) was an 11-month-old infant who was used primarily as a subject of experimentation with conditioning emotions. Watson and Raynor wanted to further the basic principles of Behaviorism by showing that any of a variety of human emotions can be elicited or removed by basic conditioning principles. For example, show most infants a small doll or toy and they will smile and reach out to it, as to them it is something positive or desirable. When Little Albert reached out to touch a white rat that was shown to him, a loud and frightening clang was made with a hammer and iron rod. This event was repeated until Little Albert began to cry even at the sight of the small white rat. Watson further continued his research with Little Albert by showing that the fear of the original stimulus could be extended or generalized to objects with a similar physical appearance (a pillow, a Santa Claus mask, a white blanket, etc.). At the conclusion of the "successful" experiment, Watson failed to address the fears that were instilled in the infant (a term that is referred to as "counter conditioning") and returned the child to the custody of his mother in an emotionally traumatized state. Additionally, more recently some research has identified that Little Albert was actually diagnosed as being hydrocephalic, thereby making his responses to the testing questionable due to his cognitive impairment. Author Tom Bartlett (2012) further suggests that Watson may have been aware that his subject was suffering from a neurological and psychological impairment, but yet continued with his own agenda of research.

RAMIFICATIONS OF BEHAVIORISM WITH EVOLUTIONARY THEORY

We have provided a brief overview of the development of how the theory of Behaviorism actually developed through the innovative and groundbreaking ideas of John Watson. Watson wanted to essentially show the world that the discipline of Behaviorism was a new scientific addition to psychology, and that those humans who were most adept in learning and adapting to their environment were most likely to survive. From many perspectives we may see how the development of Behaviorism coincided with evolutionary psychology. Darwin intended to show the scientific community how humans and all animals were capable of surviving even in the most inhospitable environments. Through genetic variation and adaptation with changing environmental conditions, humans who were able to withstand the demands of their environment not only survived, but more importantly they passed on their genetic traits (i.e., DNA) to future offspring that would further enable generations to survive in similar types of conditions.

John Watson, like Charles Darwin, wanted to present a theory that best explained human learning and adaptation based on different types of environments. Watson, like Darwin, argued that humans as a species could only survive if they were capable of understanding the environmental conditions that they lived in and were able to modify their behaviors based on their experiences within that environment. A certain type of behavior that provided success in hunting game, for example, would be repeated simply because it provided the necessities for surviving and providing resources for one's family. Conditions of the physical environment determined both phenotypical characteristics of the organism and behaviors within the environment. Those animals that failed to adapt to these changes within the environment perished.

Those individuals who were capable of adapting and changing their behaviors based on their experiences would clearly be provide with an advantage of survival. Thus, the relationship between Behaviorism and Evolutionary Psychology is dependent on the fact that environment produced essential outcomes that determined success (survival) or failure (loss of reproduction, essential resources for survival, and ultimately death).

Ivan Pavlov (1849–1936)

Ivan Pavlov was a brilliant Russian Physiologist who actually never considered himself to be a practicing psychologist or clinician. He preferred to consider himself a scientist researcher and expert in the field who explored the physiological digestive systems of mammals. He was a highly motivated student and researcher who was fascinated with the digestive systems of mammals, and described his passion to learn as his "instinct for research." Similar to William James, Pavlov was more concerned about producing the highest quality of work and was less concerned with titles, reputations or academic prestige.

Despite this professed modesty, Pavlov was awarded the Nobel Peace Prize in Medicine in 1904. Pavlov devoted his life and career to understanding and improving the world of science. He came from a highly religious background and even began his academic career at a local theological seminary, but he grew impatient with religion and decided to devote his career to science and improving our understanding of the digestive systems and physiology of mammals. Pavlov graduated from the prestigious Medical Military Academy in 1879 and continued his work studying the digestive systems of dogs with Carl Ludwig at the Heidenhain laboratories in Breslau. Here Pavlov created the now famous "Pavlov Pouch" that contained gastric fluids and began studying the relationship between portions of the stomach of dogs and the rate of salivation flows.

Perhaps Pavlov's greatest discovery in science (and subsequently psychology) was the *conditioned reflex.* Pavlov and John Watson were contemporaries and were both studying the effects of environmental factors on learning and autonomic nervous system functioning. Pavlov discovered that a previously neutral stimulus, such as the ringing of a bell, an electric shock, or a light, could be associated (i.e., "paired") with an unconditioned stimulus (UCS) (i.e., food) to produce an involuntary response (later called the "conditioned response").

The ability to predict (and ultimately control) the process of learning pending environmental manipulation was a remarkable breakthrough in psychology. The fact that specific environmental cues could predict and elicit a variety of human behaviors helped Behaviorism become the most popular branch of science in the nineteenth century. Pavlov was primarily interested in the digestive systems of animals and more specifically how the digestive glands secreted saliva at the stimulation of certain tastes and odors.

Pavlov's interest was aroused when he noticed during the initial course of testing that salivation in the mouths of dogs began actually before the presence of food. The dogs had somehow anticipated that food (UCS) was associated with other stimuli, and that this anticipation elicited a very basic unconditioned response through salivation (referred to as the UCR). This "serendipitous" form of learning was gradually referred to as classical conditioning. Pavlov then set out to further his understanding of classical conditioning by using a variety of neutral stimuli (CS) that were paired with or associated with the UCS (i.e., food or meat powder).

After a few short pairings, the dogs began to associate the food with the bell, and now simply the activation of the bell elicited the dog's salivation or conditioned response:

UCS (Food / Meat Powder) → UCR (Salivation)

Conditioned Stimulus (Bell) → Conditioned Response (Salivation)

This major breakthrough in learning and classical conditioning soon became very popular in the United States, and had a profound influence in the development of John Watson's work in Behaviorism. Soon all environmental conditions were considered to have a profound impact on learning (i.e., rewards, such as praise or punishments) and the "wave" of Behaviorism of the early nineteenth century began. Principles of classical conditioning are still used in the applied use of psychological principles, such as treating phobic conditions and cognitive behavioral interventions. Research in classical conditioning and learning was gradually associated with the work of John Garcia and the principles of learned taste aversion. The concepts of classical conditioning and learning also have had a direct influence in evolutionary psychology. Humans who were capable of identifying specific factors that contributed to their overall reproductive fitness through environmental adaptation clearly held advantages to their survival.

Humans needed to develop some form of checks and balances regarding the safety of food consumption during their evolutionary history—the practice of eating toxic foods by trial and error simply was too costly, as plants developed their forms of alkaloids and toxins that soon proved to be highly fatal to humans. The practice of consuming particular foods and associating those foods (tastes as well as scents and odors) with nausea evolved as a highly efficient mechanism in which people learned to avoid the dangerous foods by way of associating nausea (i.e., vomiting) with specific toxins and foods. By way of associating nausea as a conditioned stimulus with the poisonous food (UCS), humans quickly adapted and learned to distinguish nutritious versus toxic foods.

John Garcia, Classical Conditioning and Learned Taste Aversion as Evolutionary Adaptive Mechanisms

Many researchers credit John Garcia with the development and advancement of theories addressing learned taste aversion. Garcia (Garcia and Kimeldorf 1957) noted in his earliest works that rats exposed to radiation prior to drinking sweetend water quickly associate the nausea from the radiation and subsequently avoided the sweetened water. Since the classic works of John Garcia, much has been published in identifying key elements of the principles of learned taste aversion. Kenneth Rusiniak, John Garcia and Walter Hankins (1976) have identified several components of the evolutionary nature of learned taste aversion and how humans have adapted survival mechanisms in the process of acquiring and learning new foods. Some of the defenses include the following (*www.conditionedtasteaversion.net*):

- Novel Food Search Strategies—This particular food strategy simply limits the new or unique types of foods to those that are known to be safe. Any new or exotic type of food that has no known effect on the well-being of humans was consumed very lightly and over a significantly longer period of time. Obviously, heavy consumption of an unknown food source could have had fatal consequences;
- Increased Physiological Tolerance to Toxic Foods—Vomiting is commonly seen among many animal groups as a means of regurgitating food for offspring or distributed altruistically for other members of a pack or den (i.e., The African Wild Dog or Painted Dog, *lycaon pictus*). The capacity to expel any toxic substance after consumption was a remarkable evolutionary adaptive mechanism. Not only was the animal capable of removing the threatening substance from its digestive tract, an additional evolutionary adaptive mechanism was in the animal's ability to "learn" via biological preparedness to avoid those specific foods during future food scavenging. In several species where animals have been exposed to any type of toxic substance, after illness re-exposure to the same toxic substance typically yields heaving, nausea and ultimately vomiting. After the toxic substance is consumed by the animal, a series of life-saving neurological

and physiological mechanisms are activated. The toxins make their way through the blood stream to the medulla (part of hind brain that controls life-sustaining mechanisms such as breathing and respiration) which automatically begins the vomiting reflex. In addition to the process of vomiting, the liver detoxifies itself through the release of various toxins consumed.

- Instinctive Fears or Phobias of Unknown Foods (Neophobia)—How many children have gagged at the initial taste of a new food that is different from their more traditional foods? Broccoli, spinach, certain meats have traditionally been difficult food form to most children as they have not quite yet learned to "acquire the taste" of these exotic or different food types. Indeed, many parents report children either spitting up, gagging or vomiting after well-intentioned parents have attempted to introduce newer foods to children.

Humans have always been afraid of what they can't see or the unknown. A highly adaptive mechanism (albeit very limited to taste experimentation) was simply only consuming those foods that are known to be safe and nutritious. Experimenting with foods that appear to be attractive on the exterior (such as the bright red Baneberry) could have had fatal consequences to any human in search of food. The common tomato (*Lycopersicon Esculentum*) was considered highly poisonous as late as the sixteenth century, when they were brought to Europe from South America, and the philosopher Galen in the third century also noted that the "bright red" fruits of the Americas were toxic to eat. As a result, the tomato was a fruit that was not consumed until centuries later.

- Conditioned Taste Aversion—Regardless of how safe and cautious humans may have been in their early evolutionary history, at some point eating an unripe, toxic or spoiled food was inevitable. The physiological and psychological consequences of food poisoning are unfortunately well-known among most people, with nausea experienced shortly (10–20 minutes) after consuming the tainted food. There had to have been a highly effective and long-term effect regarding human safety that would prevent repeated episodes of consuming spoiled or toxic foods. The conditioned or learned taste aversion is an evolutionary adaptive mechanism that helps prevent humans from repeating a potentially fatal practice of consuming foods that are toxic or dangerous. Most individuals (after recuperation) who even think of the foods that made them ill initially report having no desire to eat the same foods in the future—in many cases years later. This is why in certain types of chemotherapy (where nausea is very common after exposure to radiation), doctors inform their patients not to eat any cherished or favorite foods prior to treatment, as this will prevent them from eating and enjoying those foods after radiation treatment.

In summary, without an inborn ability to detect foods that may have had lethal or detrimental consequences to our overall health and well-being, humans would have likely become extinct thousands of years ago. The ability to gradually detect and differenciate those foods that are nutritious and those that are harmful played an important role in the evolution and development of humans in general. Becoming nauseas and ill not only helped to identify potentially lethal food sources, it also helped humans to focus on the more highly nutritious foods that were suitable for consumption. These traits and practices were then practiced among future generations.

OPERANT CONDITIONING, NATURAL SELECTION, AND SELECTION BY CONSEQUENCES

An interesting link between earlier schools of Behaviorism (as described by the works of B. F. Skinner, 1938 and 1953) and evolutionary theory is the pivotal role the environment holds in maintaining (or extinguishing) human behavior. Nature "selects" those physical traits that have proven to be adaptive and beneficial in survival of the organism within specific environments, and those groups that fail to adapt will gradually become extinct. In a similar way, Skinner argues that human behavior is determined (and defined by) the various subtle reinforcing contingencies within our environment. Human behaviors emitted from the organism are functions of the dependent variables and broad ranges of differing environments become the independent variables that produce a multitude of different forms of behavior. In order to effectively understand (and predict) human behaviors we must first understand the constructs of the environment itself. For example, what are the contributing factors that promote positive and socially redeeming behaviors, such as altruism and volunteerism? Conversely, what are the environmental contributing factors that promote more destructive and violent behaviors, such as domestic violence and physical abuse? Skinner argued that a more accurate and objective assessment of the environment itself would allow us to make positive changes within our community and subsequently improve the quality of life for those subjected to inhumane conditions throughout the world.

Skinner often had much to say about improving our world and community and what we needed to do to make these improvements, such as the recommendations made in Walden II (Skinner 1948). Unfortunately, many people then (and still today) rejected Skinner's insight and refuse to acknowledge how the environment influences their behaviors that contribute to depression, substance abuse, and a deteriorating environment (i.e., pollution). Skinner became increasingly vexed late in his career at the inability or reluctance of public officials in government, education and medical establishments to realize the viability and importance of contingencies and schedules of reinforcement in changing maladaptive and destructive behaviors into productive and healthier behaviors.

The potential of helping people overcome many different forms of problems and subsequently reaching individual potential can be realized when government and communities help create systems that reinforce productive and meaningful behaviors, as described in Walden Two. Only through our ability to identify the negative environmental factors that contribute to common problems today (i.e., overpopulation, violence, war, famine) can we make more positive and effective change possible to enhance the futures of all persons in society.

The problems and maladaptive behaviors that are commonly cited by Skinner back in the early nineteenth century still persist today—conflict, war, overpopulation and a variety of health-related problems persist in light of what Pennypacker (1992) refers to as a form of "invented wisdom" where cultures support a variety of policies that are demonstrably counterproductive to ideal human functioning and adaptation. For example, within education, Skinner had demonstrated the effectiveness and utilitarianism of what he referred to as "teaching machines" to first-year high school students the basic concepts of algebra (Skinner 1984). Despite the empirically supported results (the students in the experimental group exceeded beyond traditional teaching methods and mastered twice the amount of algebra than students in the control group), the high school administrators did not implement the advanced teaching mechanisms proposed by Skinner, and the mathematical (and other school-related subjects) performance of most high school students unfortunately continues to languish today. The Program for International Student Assessment (PISA) places the United States at 29th overall regarding math skills and 22nd in science skills. Furthermore, it should be noted that these performance skills among students within the United States have consistently been dropping within the last four years, as Skinner actually predicted (Star Tribune Opinion article: "*If Only Education Were Extracurricular,*" December 30, 2013). Skinner himself noted that "A culture that is not willing to to accept scientific advances in the understanding of human behavior, together with the technology which emerges from these advances, will eventually be replaced by a culture that is" (Skinner 1984, p. 953).

Pennypacker (1992) argues that the concept of "selection by principle" (Skinner 1981) continues to evolve and influence how policies are determined in education, economics and culture itself. Rather than having *science* determine how public policy is implemented within various institutions in society (i.e., developing absolute standards for various grade levels as criteria for academic performance), *cultural practices* now define what policies are being implemented in various institutions, often with disastrous results. As Pennypacker (1992) illustrates: "Selecting one cultural practice or eliminating another changes the culture in a manner analogous to the way an individual organism is changed by exposure to a reinforcing contingency" (p. 1493). Pennypacker argues that the information and empirically verified policies defined by psychological science in operant conditioning are themselves becoming the objects of natural selection through "culturally-invented practices of wisdom" (p. 1491).

Skinner argues that the potential effectiveness of change offered through operant conditioning can only develop when communities become more actively involved in the political decision-making processes that exist within a society or democracy. In other words, Skinner knew that the only way to make contemporary society aware of the coming future problems is to *show* how the future would appear today if current maladaptive cultures continue to dictate social policy. Finally, Skinner argued that the past mistakes committed by some individuals does not shape how future decisions are made. There are many new and important issues today that require an understanding of how public policy decisions are made, including environmental issues, education and public health. Mistakes caused by human greed, negligence and aggression could be altered through a better understanding of reinforcing systems within our environment.

Part II

THE EVOLUTIONARY COMPONENTS OF NATURAL SELECTION

Chapter 5

The Evolutionary Components of Natural Selection

Despite the numerous and varied interpretations of human behaviors throughout the decades, psychology has historically remained a science that has been devoted to understanding and explaining human behaviors within a variety of different environments. Evolutionary psychology is a discipline within psychology that has grown in popularity among scientists and theoreticians. Natural selection and variation are key elements described by Darwin in support of the basic principles of evolutionary psychology. Basic principles of evolutionary psychology (i.e., culture, gender and race) have been described in a very different interpretation in recent publications (Shields and Bhatia 2009).

As we have seen throughout this text, the interpretations of the causal influences of human behaviors have varied greatly since the early philosophical interpretations and explanations of behaviors through the works of Rene Descartes and John Locke (i.e., the problem of the Mind/Duality distinction), through Watsonian behavioral learning principles (i.e., rewards and punishments), cognition and development within the environment and finally how our own evolutionary history has gradually shaped our behaviors. The Materialists of the early Greek philosophy (i.e., Empedocles) had argued that physical sciences of the world can account and explain for all components of human behavior. This rational adaptation to human behavior provides the foundation from which evolutionary psychology developed.

PHILOSOPHICAL INFLUENCES OF NATURAL SELECTION

The concept that living things can change physical structure and appearance over extended periods of time was not a revolutionary idea, even in

Darwin's time (Gribbon 2002). Earlier philosophers and naturalists had commented that an animal's structure can become modified given varying environmental conditions that impacted development over time, such as the neck sizes of giraffes increase with each successive population (see Jean Baptists Lamarck). Empedocles had argued as early as the third century that organisms and their physical structure become affected by environmental conditions such as heat and cold. After reviewing much of Darwin's work addressing evolutionary theory and natural selection, we can see an interesting influence with the broader philosophical theory of Materialism. Aristotle, for example, questions whether human development and specific characteristics of own physical body (i.e., the evolution of human teeth, skin color and eye color) were incidental visible physical developments over time or utilitarian by design. Aristotle clearly paves the way for Darwin's natural selection by arguing all things human exist for clear reasons: "Where so ever, therefore, all things together (all the parts of one whole) happened like they were made for something . . . and things not constituted perished" (*Physicae Auscultationes*, lib. 2, cap 8).

Evolutionary theory is increasingly seen as a science that has the capacity to combine and unify other disciplines into one single and logical framework (Badcock 2012). An important component of evolutionary theory, natural selection, refers to a process where certain biological traits are either increased or reduced given a particular set of environmental variables, such as climate, geographic terrains, predation, etc. This distinct range of environmental variables would gradually (over very long periods of time) determine reproductive success based on the existence of key traits that helped organisms to either survive or perish. Over several generations, the genetic variability that is common within all organisms will gradually produce phenotypical traits that make survival not only possible, but will allow the organisms within particular groups to flourish. Thus, genetic variability is the *sine qua non* of evolutionary theory and human adaptation.

Evolutionary psychology has most recently developed as a discipline that explains how adaptive behaviors and traits have evolved among individuals that have played key roles in positive styles of human interaction (i.e., cooperation) as well as more destructive or antisocial behaviors (i.e., prejudice, discrimination and conflict). The period between the late eighteenth and early nineteenth centuries was one of exciting changes and promises among those researchers in science. Leading economists were predicting dire consequences if the population continued to grow unchecked, and that the majority of humans would live in conditions unsuitable for healthy development due to problems of overpopulation (Malthus 1789).

The world was a rapidly changing place during the time Darwin was writing his 1859 manuscript: *On the Origin of Species*. Religious conservatism was very popular and Darwin knew that he would be vilified as a heretic when his evolutionary theories challenged more traditional religious theories addressing the Genesis of the human condition. When Darwin returned to England after collecting data from his journey to the Galapagos Islands, he was driven to publish his works quickly due to pressure from colleagues who pursued similar ideas (i.e., Alfred Wallace). Darwin married in 1839 and devoted all of his time to publishing and writing his evolutionary theory. He avoided social engagements, became a recluse of sorts and even complained of physical ailments (i.e., digestive problems) when forced to appear in public (Desmond 1997).

The "zeitgeist" of that era was one of breaking traditions that were currently maintained through religion and conservative policies. Psychologists, biologists and anthropologists were now suggesting that human evolution may have had occurred within the natural world and not through theories of Creationism. In the early 1850s, chimps and orangutans were on public display (i.e., zoos, carnivals and museums) and rapidly became popular objects of intrigue and fascination among the general population. People from all walks of life were now beginning to question our own evolutionary background and ancestry. The dominance of religions that had influenced perceptions of the origins of humanity was now beginning to change. People were more generally comfortable in the assumption that humans were both physically, morally and psychologically different from the primates; this view had helped distinguish humans from animals. However, as people were now viewing and studying animals (especially the great apes, monkeys and the orangutans), this convenient classification was no longer valid. People needed to understand the relationship between humans and animals, and perhaps that humans are more closely related to the primates than previously thought.

The animals that were now coming from so many exotic and different locations of the world were beginning to raise disturbing questions from a broad range of people—not just the highly educated groups of individuals who had been studying evolutionary theory. The chimps and orangutans in particular seemed to capture the interest of many people due to their "human-like" qualities. This new understanding of the common links between lower primates was somewhat disturbing for some people, because these common appearances between humans and primates strongly contradicted centuries of religious dogma and cultural belief systems. Humans were *supposed* to be both mentally superior to animals and to appear physically differently than animals. The chimps and orangutans that were now popular in zoos across Europe were strikingly similar to human behavior and appearance, as they seemed to smile, interact and even share foods with one another. The newly introduced chimps and apes made many Europeans take a second look at the origins of humanity and the growing problem between their traditional religious beliefs and the new science of evolutionary psychology.

Darwin himself visited the now famous orangutan "Jenny" at the London Zoo in 1838 on several occasions. He was intrigued with the amazing "human-like" qualities the lower primates displayed, especially when the Jenny was adorned in a child's gown as a means of further amplifying her human-like qualities. Jenny was even taught how to play small musical instruments (i.e., a harmonica) and perform in front of the patrons of the zoo where she lived. As a result of greater exposure to mammals and the lower primates, the general public was unsatisfied with traditional interpretations regarding the origins of our own species.

In a word, the world was now demanding more of a scientific interpretation explaining human behavior. Contributions from other disciplines such as anthropology (the universal expression of emotions and facial expressions), geology (the increasing discoveries of fossils), and biology (the interesting and diverse display of phenotypical characteristics among the vast variety of finches and other animals) all suggested that a new science was developing: Evolutionary Psychology.

TYPES OF NATURAL SELECTION: SEXUAL AND ENVIRONMENTAL

Natural selection was a term introduced by Darwin in 1859 to account for changes in traits among species as a direct outcome of reproductive success and interaction with a variety of environmental and ecological factors that attempted to show a common origin among humans. Darwin (1859) specifically defines natural selection as "the principle that states any variation or trait is preserved if proven useful" (p. 61). Economic and environmental conditions were critical factors relative to evolutionary theory and natural selection. As any given species tends to over-reproduce, certain environmental conditions would prove favorable to a limited number of organisms within that population. These unique variations would be preserved and less useful genetic variations would eventually become recessive and gradually extinct.

"Environmental" selection refers to the degree in which environment or ecological factors influence the degree in which specific traits are passed on to future generations (i.e., reproductive success) whereas the concept of "sexual selection" refers more to competition among members of the same species competing for child-bearing mates. During mating seasons among mammals (i.e., "ruts" for deer, elk and bear) it is very common for males among several different species to battle and compete for receptive fertile females. More recent research has now shown that the lower primates are not the only species to detect ovulating females during times of reproduction, as some cues (i.e., physical body scents, voice fluctuation and physical characteristics universally perceived as attractiveness) among human females in estrus have been detected by males (Haselton and Gildersleeve 2011; Gildersleeve, Haselton, Larson, and Pillsworth 2012).

Perhaps one of the most important points to ask of evolutionary theory is what exactly makes it distinct or different from other paradigms of psychology that attempt to explain human behavior? We have previously mentioned the importance of the cultural limitations or zeitgeist that often prevented scientists from describing their theories of the physical world. Darwin was fortunate in the sense that his timing was appropriate to present a very confrontational theory that described the origins of humanity without suffering from the recriminations from religious or politically conservative institutions. Additionally, Darwin was fortunate in having some highly influential friends and proponents of evolutionary theory who helped to defuse the controversy that evolution was now stirring among more religious and conservative groups, such as Mr. Thomas Henry Huxley (grandfather to Aldous Huxley).

The essence and overall importance relative to evolutionary theory is in identifying what suitable phenotypical (i.e., directly observable) characteristics have evolved that helped groups of animals (and plants) to survive and flourish within a specific terrain or environment and what characteristics contributed to the demise of other groups of the same animals but within different physical locations. An important factor that contributed to the ability for any organism to survive is genetic variation—without some form of variability within any species the ability to adapt to changing geographic conditions is severely limited. Genetic variation is primarily due to changes or alterations (i.e., mutations) in the karyotype structure of chromosomes which contain DNA. These alterations of genetic structure are the foundation from which evolutionary psychology is built upon.

When Darwin began taking notes during his five-year tour of the South American continent, including Tahiti, New Zealand and the Galapagos Islands in 1835, he noted that several of the species of plants and animals distributed within islands showed interesting features that seemed to facilitate their survival among the harshest of climates. The Silk Floss Tree (*Chorisia Speciosa*) grows in the tropical forests of South America and thrives especially in Brazil and Argentina. The thick and jagged spikes that project from the trunk have evolved as a highly effective mechanism in warding off pests and birds that can often damage the outer trunk of the tree, thereby making it more vulnerable to disease. Darwin noted in his epic trip of the Galapagos Islands of 1835 that animal's life itself was not alone in the direct competition for life; plants and fauna in general are also continuously adapting as a means of survival.

More importantly, however, was the fact that Darwin noted distinct differences in the physical characteristics of animals from island to island. Groups of tortoises, for example, while belonging to the same species had shown distinct discoloration of their tortoise shells from a variety of different islands. Similarly, finches (now referred to as "Darwin's Finch") also displayed interesting physical differences depending on the physical demands of different geographic locations. Darwin noted that the size and strength of the peculiar Finch bird seemed to fluctuate with changing geographical conditions of the Galapagos Islands over long periods of time. Interestingly, as the island experienced warmer atmospheric conditions such as drought, the beaks of the Finch became significantly different (harder and coarser) which Darwin attributed to an adaptation to break the shells of nuts hardened by warmer climates.

Darwin noted with interest the trend or pattern of distinct coloration that many of the native birds (i.e., finches) displayed throughout their bodies and how this colorful dispersal seemed to correlate or match much of the topography and terrain from which the birds resided. Darwin concluded that the finches (along with other several small animal groups and insects) had seemed to somehow adapt to the demands of the environment (i.e., including predators) as a means of survival. Those group members that did manage to survive and somehow escape their predators were (over long periods of time) able to pass on those specific observable or phenotypical traits to future generations, thereby increasing their overall reproductive fitness. Darwin was able to take on the herculean task of recording all available plant and animal life on the Galapagos Islands and create an efficient classification system. Gradually Darwin extended his theories of evolutionary adaptation among the lower primates to those of humans (Darwin 1959). The complete process of physical evolution is a very lengthy and gradual process, sometimes requiring millennia to make complete changes that are satisfactory with environmental demands.

Darwin further argued that the process of evolutionary change is gradual and occurs over extremely longer periods of time. Darwin realized that the manuscript (his magnum opus) he had been painstakingly working on for several years (over 20) was in peril as a friend and colleague, Alfred Wallace, was also ironically working on a very similar manuscript (without the detail and evidence, however, that Darwin had provided in his five-year travel in the Galapagos Islands). Ironically, it was Wallace who confided to Darwin several theories that were relative and similar to Darwin's own work. To add insult to injury, Wallace reported much of his findings had come to him quite serendipitously in a dream while recovering from a feverish bout of malaria.

Nonetheless, it was literally a race to the publisher to see whose work would be reviewed and accepted first. Given the fact that both evolutionary scientists has produced works that very closely resembled each other, it was determined that both men would present their work at a conference (Linnaean Society, July 1, 1858). There it was determined that a group of judges would determine which work most accurately reflected the basic components tenets of evolutionary theory. Darwin's work was selected unanimously by the judges, and that very day he sold over 1,000 copies. His new fame and fortune was unwanted and counterproductive to both his academic work and even his physical health. Darwin suffered from a series of minor physical ailments such as malaise, headaches, and fever (in addition to the premature death of his infant son). Darwin recused himself to European rehabilitation clinic for several months before he was able to return to the United States. For the remainder of his life he avoided publicity and interviews from the public. He desperately needed rehabilitation and solace from the unwanted publicity. Hereafter Darwin's name was associated with science, evolutionary theory, and his reputation became known as "The man who compared humans to gorillas."

The term "natural selection" was initially introduced by Darwin in 1859 as a means of describing how particular traits and characteristics were manifested within a species. The process was nonrandom and the scientific process was that the genome that existed in some members of a population would match characteristics of a particular environment that enabled the organism to survive. The greater the likelihood of survival, the more likely the organism would successfully reproduce and pass on these key genomes to future populations, a concept that has been referred to by Hamilton (1964) as reproductive or inclusive fitness. Individuals with greater reproductive fitness are more likely to survive and pass on adaptive traits to future populations whereas those organisms lacking in adaptive traits and characteristics are most likely to not survive. Thus a key component to natural selection is *superfecundity*, or the tendency for an organism to over-reproduce, thereby maximizing the potential for some surviving organisms to pass on key traits to future populations. Darwin further noted that natural selection was a process by which nature actively "selects" those traits or characteristics that have been determined to be most adaptive to the organism for its own survival.

It should also be noted that Darwin indicated nature does not directly affect changes in the physical appearance of an organism in his earlier works: "What natural selection cannot do is to modify the structure of one species" (*On the Origin of Species*, p. 135). This theoretically different argument was presented by Jean Baptists Lamarck in 1809 that individual animal body forms undergo physical transformations over a limited or shorter period of time (i.e., the giraffe's neck simply expands with future generations). What does occur through natural selection is the survival of those particular members of a given species who have particular traits that are well suited to a given geographic location or topography. As these particular traits have become suited to specific characteristics of the terrain, over time those genetic traits are contributed to future generations, thereby increasing likelihood of survival and increased reproductive fitness. Thus, if the basic components of natural selection are correct, then we would see a relationship between the characteristics of the physical environment and phenotypical traits. These observations were carefully recorded in Darwin's historic visit in 1835 to the Galapagos Islands on HMS *The Beagle* (i.e., wing span of Finches, tortoise shell color variation, or among humans skin pigmentation).

In his original work *On the Origin of Species* (1959), Darwin specifically notes four conditions or pre-requisites for successful natural selection to occur. The first trait necessary in natural selection is a tendency to over populate or frequent reproductive success (referred to as *superfecundity*). A second requisite of natural selection is genetic variation. Without genetic variation, groups of individual species would lack the ability to adapt to the increasingly broader demands of the environment, thus risking extinction. For example, during the Industrial Revolution in England (1819), sediment and soot became increasingly more common within urban environments. Prior to the Industrial Revolution, the white moth was considerably more advanced in population due to the characteristics of the environment (i.e., snow and light colored terrain).

The evolution of the common moth (*Biston betularia f. typical* or white moth and *Biston betularia f. carbonaria* peppered or black moth) (see Figures 5.1 and 5.2 below) has been frequently cited as a robust example of the process of natural selection over the last 200 years.

As the industrial revolution of England progressed, the physical traits and characteristics of the environment also changed radically. In particular, during the winter months the once white snow was now darkened from the soot and ashes from nearby industries. This change in physical conditions of the environment gave the darker moth (*Biston betularia f. carbonaria*) a clear

Figure 5.1 Biston betularia betularia morpha carbonaria, the black-bodied peppered moth. *Source*: Olaf Leillinger / CC-BY-SA-3.0

Figure 5.2 Biston betularia betularia morpha typica, the white-bodied peppered moth. *Source*: Olaf Leillinger / CC-BY-SA-3.0.

advantage in survival. Prior to 1800, the darker moths were very rare, but due to the radical environmental changes the darker moth successfully reproduced and increased its inclusive fitness soon over developing surpassing the lighter colored moth (*Biston betularia f. typical*). Without genetic variation among humans (i.e., differences in hair, skin and eye phenotypical variation), our ability to survive would have been threatened. Darker skin pigmentation is clearly adaptive for survival within warmer climates and lighter pigmentation more adaptive in colder climates. Changes in the physical environment through human development and industry (i.e., air pollution) had direct consequences on the advanced development of some insects (i.e., *Biston betularia f. carbonaria*) and also influenced the near extinction of other species (*Biston betularia f. typical*).

The third requirement for natural selection to occur is inheritance. Our ability to survive depends not only on reproduction, but more importantly our ability as a species to pass on traits to future generations that have been proven to be adaptive within specific types of environments. Finally, populations within a species must have variance of survival and reproductive traits. As populations continue to grow, individuals must be able to adapt to different types of resources that are made allowable within their environment. In other words, a specific ratio of survival and mortality must exist and a specific environment is limited to support a limited number of individuals within any group. When the population exceeds this finite number, nature will "weed out" or eliminate those individuals not possessing traits that are most suited for surviving.

The term "inclusive fitness" has been offered by numerous physical anthropologists (i.e., Hamilton 1964; Dawkins 1976) as a means of describing how various organisms may improve their reproductive and genetic success through social and collaborative (even prosocial and altruistic) behaviors with other organisms. Hamilton (1964) has even identified a mathematical model of inclusive fitness where a gene can increase its reproductive success by promoting the development of genetically related individuals or kin (sometimes referred to as kin selection): $r\text{B} > \text{C}$, where r refers to the correlation of the average number or amount of genes shared with B (benefits received from others who are likely to reproduce) is greater than the overall cost (C) of the altruistic act.

In conclusion, Darwin's theory of evolution and the topic of natural selection provide an interesting view of how the environment interacts with biology and reproductive fitness. According to Charles Darwin, nature gradually selects those physical traits and characteristics that help make survival most feasible and possible. Animals that were most likely able to camouflage and blend with the physical characteristics of the environment were significantly more likely to survive and reproduce, thereby passing on these traits to future populations.

The single most important characteristic to evolutionary psychology (and ultimately our own survival as a species) is that of *genetic variation*. Without some means genetic variability and adaptation, humans could not have survived in an environment that placed so many demands on its inhabitants. Those species that were capable of adapting to constantly changing environmental circumstances, involving climate change and availability of food as a key resource were most likely to pass on those physical traits and characteristics to future populations, thereby ensuring a complementary relationship between host and environment.

Both Darwin's contribution of natural selection in 1859 (*On the Origin of Species*) and the profound scientific reasoning of the classic philosophical theories as described by the Materialists, Positivists, and Empiricists agreed in one fundamental principle: All interpretations of human behavior would be a direct function and consequence of a scientific understanding of natural and physical forces in the world. Additionally, these forces could be traced to much smaller and indivisible elements that we now refer to as "atoms." There was no longer any room for metaphysical or divine interpretations of behaviors—all things physical defined the rational and logical world. The zeitgeist of European contributions of the nineteenth century was now increasingly rational and pragmatic, and the once popular metaphysical belief systems were being replaced with a more observable and measurable science that we refer to as evolutionary psychology.

The earliest of the Materialists, Epicurus (341–270 BC) and Darwin both faced public criticism and even persecution for challenging the metaphysical and profound religious doctrines and dogma of their eras. As early as the 340 BC, Epicurus argued that the basis of all knowledge should be observable and measurable events within our world and logical deduction. Simply an inability to explain events scientifically, according to Epicurus did not justify the use of "spiritual" explanations or metaphysical explanations of events occurring within our world. Additionally, Epicurus supported the reasoning that all matter was comprised of small particles of matter flying through the atmosphere or "kenos." These smaller and indivisible bits of matter later came to be known as atomic structures and provided the framework from which all organisms develop from.

The early Greek philosophers and Darwin fundamentally agreed that events occurring within the physical world can (and should) be explained through a scientific understanding of the basic elements of nature. Neither Epicurus nor Darwin were self-proclaimed atheists (Darwin himself was an agnostic), as Epicurus did participate in traditional religious ceremonies and Darwin attended a Church of England school and even considered the clergy as a profession during his younger years while at the University of Cambridge. Darwin (1887) claimed that "Science has nothing to do with

Christ . . . as for myself; I believe that every man must judge for himself between conflicting vague probabilities" (pp. 304–307). Thus, it is important to note here that as earlier as the 340 BC with Epicurus through the nineteenth-century science was now taking a more prominent position in terms of education within the general public and understanding the events of the physical world. Both Epicurus and Darwin attempted to prove that science and religion were not mutually exclusive (at least in academia) and that the world's most foremost authority on evolution did not reject the idea of God.

While no outright rejection of a "Superior Being" ever existed in any of Charles Darwin's works, it is clear that the "immense suffering of the world" had caused Darwin to reflect on the feasibility of the existence of a God. The untimely death of his beloved and favorite daughter Annie (April 23, 1851) also seemed to contribute to Darwin's growing skepticism addressing the existence of a God in light of universal pain, suffering and oppression (Darwin 1887). As his work on evolutionary theory became increasingly developed and advanced, Darwin accordingly grew increasingly vague and critical regarding the existence of Divine Power or the existence of "miracles." Given Darwin's extensive background and early training in Anglican Theology as a young man in England he appears dutifully obligated to remain at least quasi-theistic. In an interesting summary describing Darwin's views on religion, Sara Joan Miles (2001) notes that Darwin is troubled by this topic and writes a letter to his good friend and American collaborator Asa Gray:

> "With respect to the question of theology . . . I am bewildered. I have had no intention to write atheistically . . . there seems too much misery in the world and I cannot persuade myself that a beneficent and omnipotent God would have designedly created the *lichneumonidae* with the express intention of feeding within living bodies of caterpillars. I am inclined to look at everything resulting from designed laws [of nature] . . . I feel most deeply that the whole subject is too profound for human intellect—let each man hope and believe what he can (p. 197)."

In conclusion, it appears that Darwin's convictions regarding the existence of God were highly complicated—it seems as though he desperately wanted to believe in some form or type of God, but the growing body of research in his own work (including his personal grief with the premature death of his daughter) limited his ability to confirm the existence of a spiritual deity. Indeed, Darwin stopped attending formal church services with his own family in 1849, and he later confided to a friend John Fordyce in 1879 that he could best be described as an "agnostic." More importantly, he further confided to his friend that: "We can be an ardent theist and an evolutionist" (Darwin letter to John Fordyce, May 7, 1979).

THE EVOLUTION OF COOPERATION: THE "GLUE" THAT MAINTAINS SOLIDARITY

Much of what we have been describing previously about the history of the development of evolutionary psychology has addressed early philosophy of human behavior and Darwin's theory describing how the environment influences the physical development and appearance of organisms in their efforts to reproduce and survive. Darwin's classic evolutionary theories at first glance seem to make strong arguments for the virtues of self-gain over sacrifices for group development. Similarly, traditional Darwinian Theory seems to make a strong argument that all behaviors that produce benefits exclusively to individuals and outcompeting rivals for resources would maximize reproductive fitness and progeny.

Stereotypical views of early human nature as being greedy opportunists living in a "dog eat dog" world are commonly held by many people—scholars and laypersons alike. However, such is not the case, as a closer view of human nature during our evolutionary past shows a much different picture. For example, we know that in several disciplines such as psychology, sociology, anthropology, and even organic chemistry that sacrifices that are made by individuals can not only (in the long run) provide advantages to the group in which one resides, but also can maximize benefits to the individual making the sacrifice to the group itself. However, another, perhaps even more important component of evolution addresses not necessarily how human species changed physically in their never-ending efforts to survive, but rather in how their relationships with each and levels of cooperation among one another also changed as a means of ensuring survival.

Humans clearly are not the only species to exhibit and participate in a variety of cooperative behaviors. Vampire bats, despite their reputation as blood-thirsty and ugly creatures, have in fact displayed reciprocally cooperative and altruistic behaviors in groups for centuries. When a vampire bat is not successful in securing food from prey, the bat will solicit food from other bats within the same colony and will typically be given regurgitated food by a nonrelative. The sacrificing bat "remembers" which of his peers he has altruistically shared his meal with and in times of need he will receive food from his previous partner. This behavior is very similar to humans who "owe" each other favors and are more likely to help those who have helped us in the past. The overall health of the group is actually increased and becomes more versatile when members are willing to make sacrifices for individuals in time of need.

Additionally, current research suggests that prosocial behaviors may not only be learned during early phases of development but more importantly are now considered reflexive and intuitive behaviors that were necessary as groups evolved and engaged in social interaction (Zaki and Mitchell 2013). While many developmental and evolutionary psychologists recognize the significance of prosocial behaviors among humans during early evolutionary periods of developments (Warneken and Tomasillo 2009), the origins of such behaviors remain a highly controversial topic. Recent research suggests that prosocial behaviors were necessary forms of engagement during social interaction during our evolutionary history (Hamlin, Wynn, and Bloom 2007).

Cooperation and reciprocity were vexing problems for Darwin's theory of evolution: These concepts were not perceived to be possible in a world containing the antisocial and aggressive types commonly described in early history. Primates, such as the chimps, gorillas and orangutans have been observed to present "gifts" or "nuptials" to each other to facilitate rituals such as courtship, mating and general social interaction. Even insects, such as the common beetle (*Calosobruchus macolatus*) share vital resources such as water to facilitate a "quid pro quo" process of reproduction—males share liquids such as water to females who need the water while manufacturing ova, and in the process of providing water males deposit their sperm for reproductive purposes (*New York Times*, 12/23/13).

Biology is not necessarily our destiny and life in the extreme does not always represent my gain is your loss (a zero-sum relationship). The concepts of cooperation and altruism have existed despite the egoistic and opportunistic tendencies among humans throughout evolutionary history, and more recently scholars have attempted to explain how human virtues such as altruism and cooperation could have existed (and in some cases even thrived) in a highly competitive and harsh evolutionary environment.

Martin Nowak (2012) has offered several plausible reasons why positive human traits such as cooperation, prosocial behaviors and altruism have continued to exist despite a seemingly harsh and competitive world. First Nowak argues that how cooperators and defectors are *distributed* throughout a community has much to do with prevalence and existence of cooperative behaviors. When cooperative individuals coexist in smaller groups or cluster, their ability to display helpful behaviors significantly increases. Empirical research that explores the prevalence of cooperation and altruism within a mixed group of individuals (including opportunists and defectors), the cooperators thrived despite occasional defection among others in the group.

A second factor that explains the resiliency of cooperators among defectors as described by Novak has to do with a concept referred to as kin selection. Kin selection has to do with the likelihood of helping others who may be directly or indirectly related to us, and when we do cooperate and make sacrifices for our kin, our own genetic reproductive fitness is also enhanced. Finally, Nowak describes group selection as an important factor that has contributed to the development of cooperation throughout our evolutionary history. Simply stated, when groups of individuals make sacrifices to the larger community or common good, the overall conditions of the group are improved and thereby increase how well members within the group are able to survive and face challenges.

Taken together, the benefits and importance of cooperation relative to humans are vital to our very existence, nor are humans alone in the manifestation of these traits. When we make sacrifices for others, whether the beneficiaries are kin or strangers, we are creating a more supportive and resilient community for all individuals to inhabit and exist in.

THE "PRISONER'S DILEMMA": HOW COOPERATION COMES AT A COST, BUT ENSURES MUTUAL BENEFITS

The classic Prisoner's Dilemma was originally described by Merrill Flood and Melvin Dresher (de Herdt 2003) while working at a popular think tank (RAND). The two men argued that individual profit or gain always supersedes cooperation or collaboration among two or more individuals. However, if this theory were actually true, then any form of self-sacrificial behaviors such as altruism or prosocial behavior would have been nonexistent and limited the resiliency in the development and growth of the human population.

Human cooperation has been considered to be essential to the development of productive and human behaviors in society. In our earlier philosophical discussions addressing human behaviors, Thomas Hobbes argued that cooperation is a necessary but not sufficient behavior in the development and expansion of society. Cooperation, according to Hobbes, is the antithesis of human nature as humans by nature tend to be opportunistic and greedy. Hobbes does agree, however, that a larger societal force through government can encourage individuals to engage in more cooperative behavior.

The classic Prisoner's Dilemma problem serves as a useful example in supporting Hobbes' argument that humans are biologically predisposed to greed but can and often become reluctant cooperative partners. Typically when both individuals are shown how they *both* can benefit from trust and cooperation in a situation that can have serious consequences, they are more likely to benefit by working collaboratively with each other. Conversely, if both individuals behave within their own self-interest in the Prisoner's Dilemma—then both lose and serve a longer prison sentence. However, if (prior to apprehension) the players can agree to protect each other (i.e., cooperate with each other), they will serve a shorter sentence. But the crux of this dilemma is that they need to rely and trust (i.e., develop interdependence) with each other in order to make this pact work. Thus cooperative behaviors (albeit reluctant cooperation) within our society serve the greater good if all partners abide by common rules governing their behavior.

In 1964, W. D. Hamilton's theory of inclusive fitness argues that those individuals who share close genetic structure can significantly increase reproductive fitness in offspring and relatives. His mathematical formula supports the theory that when closely related individuals share risks in the development of kin and their offspring, they are significantly more likely to pass on their genetic traits to future populations: $rb > c$, where: r = probability of an individual sharing an altruistic gene; b = reproductive benefits of the recipient of the altruistic act; and c = overall cost to the altruist or individual sacrificial behavior.

In the classic Prisoner's Dilemma, two criminals are apprehended and both face a stiff penalty (20 years) if the district attorney can acquire more information from either criminal. As a means of soliciting more testimony, a "deal" is offered to one of the prisoners in that if he or she confesses the information (i.e., "squeals"). The dilemma occurs when two people know that if they cooperate with each other they will get reduced sentences (one year), but if one confesses their crime; their partner will receive the maximum penalty (20 years).

The Prisoner's Dilemma has important implications in sociology, psychology and evolutionary psychology. The ability to forge and maintain relationships with other (non-related) humans required trust and cooperation. Without it, humans clearly could not have survived the numerous challenges that existed relative to their well-being and health. Food, shelter, and protection all required humans to work collectively together for the common good. Imagine that you are now a Neanderthal person over 100,000 years ago and you were in constant search of calories (i.e., high protein food sources, such as deer). You know where healthy deer graze, but you needed other people to help you trap and harvest your potential food source. True—if you could kill your food source by yourself you would have significantly more food without the need to share. However, trapping and harvesting deer remains a very difficult task if your only weapon is a spear or a rock. So you compromise and solicit help from other clans to acquire food resources. You share your food and in the process your hunting party exchange ideas, information and skills with you so in the long run you are more resourceful in survival skills.

This process of utilizing shared resources required cooperation and helped to forge productive relationships in an environment that was extremely hostile and difficult to survive. Cooperative behaviors not only established trust and positive forms of collaboration, it also identified how interdependency became vital in the development of small groups. Interdependency simply refers to the awareness that achieving your goal is contingent on the group achieving its goal. If everyone works together, then goals are realized and the group benefits. If, however, an opportunist does not work and collaborate relatively equally, then the group fails (and all individuals who comprise the group) to achieve its goal.

Collaborative work and shared responsibilities were therefore a proven strategy that significantly improved the resilience among humans in surviving extremely harsh environmental conditions. Many psychologists, biologists and anthropologists have argued that even though environmental conditions have radically changed since our early evolutionary history, the need for people to share their skills and expertise in a collaborative fashion still exists even today.

In short, humans learned to acquire the vital skill of cooperation not only as a means of survival but more importantly as a method of improving our communication, relationships and interaction with others. While food played a vital short-term resource, establishing collaborative processes by way of cooperation was also a vital development in human evolution over the long run. Cooperation with other non-kin members typically meant one of three things:

- Increased reproductive fitness through access to more resources. By increased interaction with other clans, information was exchanged and in some cases fertile females which increased genetic resilience among different clans; and
- Improvement in access resources, such as food and hunting techniques and acquiring essential things for survival;
- If cooperation with other groups did not exist, then isolation would result thereby reducing the clan or group's ability to survive in difficult environments.

Numerous empirical data suggest that groups of individuals experience a significantly improved way of life (i.e., psychological states of well-being) when there is access to interactive programs, community engagement opportunities, and simply ways that afford individuals opportunities to communicate and better understand one another (Sarason 1974). Humans have relied on cooperation early on in our evolutionary history as a way of survival; without some form or method of collaboration our ability to withstand negative environments (i.e., drought, sickness and illness) would have become significantly compromised.

Group work and interdependency have long provided the glue that helps different cultures to improve their relationships with one another. Community group activities have their roots in our evolutionary history, as early humans learned how vital it was to share information with each other as well as sharing in vital resources, such as food and water.

TEACHING AND PROMOTING THE DEVELOPMENT OF COOPERATIVE BEHAVIORS

Most educators today would agree that the benefits of cooperation, however salient in human interaction, are principles that can be taught to different groups of individuals, especially children and adolescents. For example, group work in classrooms in primary and secondary education that involve collaborative learning principles have been noted to improve retention and academic performance.

Axelrod (1984) notes several necessary principles in the development of cooperative behaviors among different groups of individuals:

- What Can We All Get if We Cooperate? A great motivator of behaviors has always been incentives—showing people how they may benefit when they are capable of working collaboratively with others. Axelrod notes that the most important strategy in enhanced cooperation is to simply show people what their potential rewards may be if they do participate and cooperate with others. Identify the positive aspects of group and future benefits that all group members may develop. In other words, when all neighborhood members work toward creating a safer and cleaner neighborhood, they can all enjoy an improved and safer environment in which to live;
- Change Payoffs. Axelrod notes that one of the primary reasons why some individuals refuse to cooperate with others may primarily be due to egoistic and opportunistic means—in other words, people who think that they can manipulate governmental agencies to their benefit (i.e., through embezzlement or theft) most probably will do so. Axelrod notes that in order to stop these counterproductive behaviors to cooperation, show how the payoff (i.e., money) is not worth the risk (i.e., prison time);
- Teaching People to Care for One Another. The concepts of cooperation, sharing and interdependency are not innate—they refer to a set of behaviors that must be learned. While it is true that humans have evolved with the capacity to share and cooperate, our own existence today is evidence that cooperation must have occurred that enabled groups, clans and societies to grow and evolve. No one single person can do everything, and when groups formed and merged together, individuals possessing various skills must have shared their expertise in the goal of survival and growth. The "Tit-for-Tat" principle that proved so fundamentally important in the Prisoner's Dilemma (Axelrod 1984) simply identifies the role of reciprocity—whether it is positive reciprocity (you helped me so now I am going to help you—"The Golden Rule") or negative reciprocity ("An Eye For an Eye"). If anything, showing individuals that their behaviors within a community

or society typically will have an impact on those living around them will influence their behaviors in some way.

The question remains, however, at what point in development are these skills capable of being learned, and does an actual "critical period" of learning exist relative to the development of cooperative skills? Teaching children the benefits of cooperation and shared responsibility is especially important in those cultures where competition and winning are highly valued (i.e., the individualistic culture). Parents and community members that involve children in a variety of projects that are designed to improve community development (i.e., developing a community garden, building homes for habitat, etc.) are excellent ways to teaching children the virtues of community stewardship and volunteerism.

- Teaching Reciprocity—How "Tit-for-Tat" Can Enhance Interdependency in Cooperative Group Work

Developmental psychologists are aware that all children, regardless of cultural identity and backgrounds, pass through "sensitive" or "critical" periods of time when they are more capable and receptive to learning basic skills. Children can learn to cooperate and share basic skills with one another and are more likely to do so with others when they are exposed to culturally diverse environments where they are encouraged to share and communicate with other children from different backgrounds. Communities need to provide opportunities for members to engage and share in community service activities with one another—regardless of the ages of the participants. When communities provide opportunities for all members to share experiences that benefit each participant (i.e., community service work projects), a sense of psychological connectedness and cohesion develop that actually strengthens the rapport and development within that community (Markus and Kitayama 1991).

Krueger, Fisher, and Wright (2013) have developed an intriguing theory that explains the motive of cooperation from the concept of "social projection": Individuals frequently cooperate only after they evaluate the probability that other's behaviors will correspond with and match their own behavior. Krueger's theory of social projection sounds very similar to Axelrod's concept of "Tit-for-Tat": A primary cognitive tool used by many individuals in whether or not to cooperate lies in our own perception if our behavior will be reciprocated. Krueger noted that while it is true that many individuals engage in a "purely altruistic" action, the majority of people engage in prosocial or helpful actions after careful consideration that the behavior will be returned in some way. This concept has been described more recently as "conditional altruism" or reciprocal prosocial behavior. Krueger also notes that cooperation typically occurs under three conditions:

a. People who cooperate do so only because they have not fully considered all consequences and costs of their behaviors (a term he refers to as "stupidity");
b. A second motive to cooperation addresses the concept of moral thinking and "conditional altruism." What this means is that the majority of individuals are most likely to cooperate (and engage in altruism) only if they believe that their cooperation or altruism will in some way be returned either to their directly or more generally to the public; and
c. Inductive reasoning: People who cooperate do so as a means of promoting their own self-interest. Similar to Robert Triver's (1971) concept of reciprocal altruism, Krueger argues through inductive reasoning that often when we help other individuals we are also increasing the likelihood of receiving help for ourselves. While the motive may be less altruistic, inductive reasoning offers a practical alternative (i.e., "quid pro quo") approach that explains why individuals may be more compelled to cooperate and exhibit prosocial behaviors within a variety of different types of environments.

An important component to the development of cooperative behaviors is in allowing individuals to understand how the concept can benefit all participants. A key element to cooperation is in allowing all participants to see that the rewards are mutual and that the benefits for all persons are relatively equal and tangible. Similarly, in order for cooperation to be taught effectively we need to not only identify the mutual benefits and rewards for all participating individuals, but we also need to illustrate the goals are interdependent and only possible through relatively equal cooperative involvement in the community project. These goals (referred to as superordinate goals) can enhance cooperative behaviors by showing people that they are only attainable when all group and community members participate (i.e., no "social loafers").

An additional benefit to the development of cooperative group work is the enhanced perception of the successful ability to achieve goals—what Bandura refers to as "collective self-efficacy" (Bandura, Fernandez-Ballesteros, Diez-Nicolas, Caprara, and Barbaranelli 2002). The concept of collective self-efficacy is essential to cooperative group work and community service work activities that involve various groups of individuals because collective self-efficacy influences how the group perceives their ability to achieve goals *collectively*. The key difference between collective self-efficacy and efficaciousness is that collective self-efficacy influences how the group perceives and ultimately completes a variety of community or group projects.

Individuals cannot complete extensive or broad projects by themselves; they need assistance from other committed individuals as a means to complete each project.

Collective self-efficacy is also relevant to community development because individuals feel significantly more empowered when working with others who have similar values and incentives to achieve the goal itself. In short, collective self-efficacy is very motivating and rewarding to experience because of the opportunity to work with others, share similar values and complete civic or community-related projects which otherwise would have remained impossible to achieve.

Finally, teaching individuals the benefits of cooperation may best be demonstrated when we ourselves are capable of a few basic rules. These rules support the simple concept of the "Tit-for-Tat" previously described that provides the most productive and cooperative strategy when two (or more) individuals are faced with a choice of maximizing personal outcomes (i.e., defection to a partner) or making limited sacrifices to ensure positive outcomes for both parties (i.e., mutual cooperation):

a. Always (at least initially) be friendly with those you intend to work with. People are far more likely to reciprocate prosocial or positive behaviors than destructive behaviors;
b. Don't be taken advantage of! If someone (i.e., your partner) does defect and in some way tries to take advantage of you or your situation, then behave in a similar way or it will happen again. This is also very similar to what many refer to as "The Golden Rule": Return positive behavior with positive behavior and if your partner defects, then you defect (i.e., "An eye for an eye");
c. Be fair to your partner (i.e., non-envious); and
d. Keep your behaviors and interactions with others simple and direct.

COMMUNITY SERVICE WORK AND COOPERATION: THE EVOLUTIONARY NEED TO COLLABORATE

Humans are not only social creatures, but we have evolved with the need to cooperate and interact with one another, sharing our skills within a variety of ways that helps improve overall conditions of living. Humans could not have existed without learning and adapting to various situations that necessitated our capacity to work as interactive and social creatures, and this very deep-seated need still exists today. Communities need to provide citizens and inhabitants with a variety of opportunities to work collaboratively together as a means of improved understanding of each other and thus to avoid potential conflicts. A theory that identifies a central causal factor of ethnic conflict and discrimination is ethnic polarization and the "We—They" dichotomy.

According to this theory, the underlying foundation of ethnic conflict is polarization (usually caused by economic and educational achievement gaps), where the dominant group views the marginalized group as deficient or inferior in a variety of ways, often blaming them for their own misfortunes (i.e., "blaming the victim"). Communities that provide opportunities for members to work together in the development of healthier and safer environments have numerous positive psychological and social benefits. Exposure to and collaboration among ethnically diverse groups to a mutually beneficial goal for all community members helps each person to discover the similarities that they have with each other and not perceived differences.

For example, when individuals from ethnically diverse environments are provided with opportunities to create and maintain a community garden and fruit tree orchard, reports of ethnocentrism significantly decreases (Hoffman, Wallach, Espinoza- Parker, and Sanchez 2009). An increasing problem today in many communities is the rapid deterioration of community projects and opportunities for people to work collaboratively, due to increased modern technology. Community service work opportunities provide individuals with opportunities to share their skills and practices within a variety of ways to promote and build stronger communities—this practice has a strong evolutionary background as early humans needed to band together and share their skills as a means of survival. Those communities that do provide ample community service work opportunities has been shown to be more resilient against crime and have more resources that are available to more community members.

FACIAL EXPRESSIONS AND EVOLUTIONARY THEORY

All facial expressions involve the manipulation of muscles beneath the skin of the face. Research strongly supports the universality of facial expressions (i.e., Ekman, Friesen, and Ellsworth 1972) and that emotions are commonly associated with facial-related expressions, such as fear and anxiety. Darwin also noted that facial expressions and morphology among humans are strikingly similar among several lower primates, notably the chimpanzees and gorillas (Vick, Waller, Parr, Smith-Pasqualini, and Bard 2006). Darwin further noted that both animals and humans were capable of expressing emotions via facial expressions and these facial expressions were vital tools used in establishing cooperative relationships. The fact that specific human facial expressions are similar to those of lower primates supports the theory that emotions are commonly displayed through a variety of facial expressions and was a critical feature in defining the development of socialization among humans (and infants).

The capacity to determine if a stranger approaches you with a facial expression demonstrating anger or fear may have profound influences in how you may approach and engage with that particular individual. Additionally, past research has shown that infants as young as three months of age are capable of discrimination (as measured in eye gaze or "fixation") between "happy" and "surprise" faces and to a lesser degree "sad" and "surprise" faces (Young-Browne, Rosenfeld, and Horowitz 1977). Darwin further noted that an animal's ability to communicate feelings by way of facial expressions was a key trait that helped improve reproductive fitness among a variety of species.

An important skill dating back to early evolutionary development was an accurate assessment of the different types of facial expressions expressed by strangers. Direct and narrow eye contact may have indicated danger whereas a smiling face complemented with wider eyes may have reflected a more welcoming approach when meeting people for the first time. Facial expressions are complicated things. They involve a series of muscles that impact the forehead, mouth and eyes. Even blinking can influence what we are feeling, as increased blinking rates are typically indicators of anxiety and deception (Leal and Vrij 2008).

An interesting topic that we will consider is the function and role that facial expressions have had through human evolution. The explanations of facial expressions from an adaptive view (i.e., facial expressions as accurate indicators of emotional affect) were first addressed by Darwin in his now famous text: "*The Expression of the Emotions in Man and Animals*" (1872).

The important findings in this text identify human emotions and facial expressions as direct tools that were used among humans as a means of establishing trust and cooperation in early development of clans as well as protection among perceived strangers. Similarly, these basic facial expressions were considered to be reflexive and instinctual tools that were used in protecting kin from perceived threats as well as developing positive and cooperative relationships. For example, a furrowed brow among the human species is relatively easy to identify and a very clear and obvious indicator of fear or anger. This emotion is considered to be universal and displayed among various cultures throughout the world (Ekman and Friesen 1986; Ekman and Keltner 1997).

The idea that facial expressions can express different forms and types of communication is not new, as Darwin described the utilitarian and distinct features of facial expressions as a function of evolution and survival as early as 1872 (*The Expression of the Emotions in Man and Animals*). The fact that distinct facial expressions can help communicate specific emotions is central to the theories of evolution and how humans have coexisted (i.e., displays of dominance over submission) with each other. The capacity to display fear, such as facial descriptions depicted with round eyes and younger or more "babyish" facial structures has been shown to elicit prosocial or helpful responses in group situations (Marsh, Kozak, and Ambady 2007). Expressive facial features and other types of facial behaviors have also been identified as instrumental in eliciting social attraction (Boone and Buck 2003). For example, seeing an expressive "happy face" may serve as an effective signal that the target face is capable of helpful and cooperative behaviors and an accurate representation of an individual's personality trait (Borkenan and Liebler 2005). Thus, becoming attracted to individuals displaying positive facial expressions such as smiling and "happy faces" can serve a highly adaptive function as they are perceived to be both cooperative and socially fit individuals. Similarly, research addressing facial expression among children found that those children with more expressive facial features where more popular in primary grades and kindergarten levels (Buck 1975).

More recently, facial expressions have also been used to accurately identify specific traits among individuals and thus serve as an effective tool in the development of social communication (Parkinson 2005). As research becomes more specific in identifying various types of facial expressions as cues for specific reactions from threatening targets, two specific facial expressions (fear and anger) have been presented by Sacco and Hugenberg (2009) as inherently unique communicators of vital information that played a pivotal role in shaping our relationships with others during evolutionary history.

In a series of three different experiments involving manipulations of younger or "babyish" (wider eye) facial expressions and mature (i.e., smaller eye) facial expressions, Sacco and Hugenberg demonstrated that specific facial characteristics evolved essentially as a mechanism that facilitated social communication. A facial expression capable of representing fear (wider eyes) could hold tremendous adaptive value in that it can help to defuse and de-escalate potentially violent situations. People who are perceived to be afraid of others are generally not perceived to be a viable threat and thus violence is significantly less likely to manifest itself in small group settings. Our ability to detect anger through mature facial features (i.e., smaller eyes) also holds tremendous adaptive value from an evolutionary perspective. As our eyes widen when we experience fear, we are better able to visually detect potential threats (i.e., increased broader visual patterns) as well as serve as cues to others (seeing the "whites of the eyes" or sclera) that a potential threat is within proximity to the group.

While children and adults who display greater amounts of facial expressions are perceived to be more socially attractive (i.e., cooperative and trustworthy), the reverse is also true for those who lack in facial expressiveness or display negative facial features (i.e., frowns) (Hubard 2001). More transparent and visual positive facial expressions appear to convey numerous advantages in social communication that serve as effective signals in adaptiveness and reproductive fitness. A frequently frowning person signals typically depression and unhappiness within that person, and subsequently an inability to establish and support environments that promote healthier standards of mental health and living (Cole 1997).

Understanding facial expressions and more specifically wide-eyed versus narrow-eyed expressions of strangers can also facilitate relationships with outside group members. For example, knowing how to selectively identify those individuals who are willing and capable of helping us in times of distress (i.e., those who are not angry) could hold tremendous advantage in our well-being and survival ability. Similarly, a highly evolutionary adaptive trait in detecting anger could not only help us to avoid potentially dangerous situations (i.e., physical altercations), but could also help in the development of cooperative relationships with others. Indeed, knowing *when* to ask for help (i.e., who is or is not angry) is often just as important in knowing *who* to ask for help to ask for help.

Most cultural psychologists and anthropologists agree that the development of facial characteristics have evolved over long periods of time and are cross-cultural (Ekman and Friesen 1986). Understanding how basic facial characteristics may have influenced how early humans forged different types of relationships with each other (i.e., smiling may have influenced social group gatherings, whereas a snarl or frown with furrowed brows may have signaled others to stay away) is a central goal of this chapter.

Given this short review of inclusive fitness and natural selection, how then have facial expressions evolved to be an adaptive mechanism in human interaction and society? Facial expressions have been described as variant forms of "nonverbal communication" (Izard 1991). Facial expressions have also been described as key indicators in human interaction that convey our thoughts, feelings, and moods with universal consistency (Ekman 2001). More recently, facial expressions have been identified as valid determinants of other characteristics including honesty (Tyler, Feldman, and Reichert 2006) and hostility (Ellsworth and Carlsmith 1973).

Historically, there have been six identified facial expressions that are considered universal (i.e., a smile as an indicator of happiness cross-culturally): disgust, fear, joy, sadness, anger, and surprise (Izard 1991). Darwin noted that the reasons why universal facial expressions have developed cross-culturally were due to their adaptive nature in displaying authentic human emotions within a variety of different types of environments. When feeling threatened, for example, humans typically display a negative facial expression as a means of warning to strangers (Greenbaum and Rosenfield 1978).

The display as well as the social awareness of understanding what these facial expressions represented was absolutely essential for survival during our evolutionary history. Without these skills, humans clearly would have become extinct years ago (Krumhuber, Manstead, Cosker, Marshall, Kappas, and Rosin 2007).

While cultural differences do exist with respect to the display of facial expressions and emotions, it is reasonable to conclude that accurately detecting facial expressions (i.e., happiness or anger) were particularly important skills that helped humans in their overall task of survival, including forming trusting relationships, cooperation and displaying aggressive features, such as a frown or direct eye contact (Elfenbein and Ambady 2002). Nonverbal communication through the development of facial expressions (and more recently "microexpressions") is considered by many social psychologists and anthropologists to be an important tool in understanding the basic emotions of individuals who are familiar to us as well as those who may be considered strangers. "Microexpressions" have been recently identified as subtle yet important smaller facial expressions that are key indicators of what people are truly feeling or expressing (Baron, Branscombe, and Byrne 2008). For example, individuals who purse or tighten their lips or quickly display a variety of different facial expressions may be displaying characteristics that have been associated with deceit or dishonesty (DePaulo, Lindsay, Malone, Muhlenbruck, Chandler, and Cooper 2003).

Behaviors that have historically been associated with positive affect, such as trust and cooperation have been associated with the directly observable facial expressions of smiles and open-arm gestures. In other words, not only a smile was considered to be an important tool in establishing trust and cooperation among our early ancestors, but also the physical position and gestures of our bodies (i.e., open arms as a universal welcome gesture and crossed arms as an indicator of anger or contempt) when confronted with strangers (Burt 1995).

More recently, anthropologists as well as evolutionary psychologists have studied facial expressions as key behaviors that served to improve our ability to survive among when exposed to unfamiliar individuals with groups (Cosmides, Tooby, and Barkow 1992). Most researchers today agree that universal facial expressions have historically served as indicators of various emotions (i.e., frowns indicating displeasure or smiles indicating pleasure) and were vital in several aspects of human interaction (Shaver, Murdaya, and Fraley 2001). Arguably, even today many individuals develop a variety of positive relationships (personal or professional) with others based on their own evaluation and impression of the subjective nature of facial expressions. Empirical research has supported the hypotheses that our perception and sensitivity to facial expressions has a significant influence in other important areas of our lives, such as communication, physical attraction, and the development of cooperative and collaborative relationships (Bradbury and Vehencamp 1998).

While the existence and development of facial expression have important evolutionary implications (Sacco and Hugenberg 2009), perceiving and understanding what facial expressions actually mean when confronted in groups also has enormous value in our ability to understand adaptive behaviors and evolutionary psychology. If Cicero is correct and "the eyes are the window to the soul", then perhaps an important component to fostering social relationships is through our ability to understand communication with others and identify the characteristics that define honesty or deception. Unfortunately, the current empirical data does not support the hypothesis that individuals are capable of detecting honesty or dishonesty in social interactions (Ekman and O'Sullivan 1991; Ekman, O'Sullivan, and Frank 1999).

However, while it may be true that no inherent gender differences exist in detecting deception among men and women's styles of communication, a growing body of research does suggest that women show greater accuracy in identifying moods that are typically reflected through facial expressions. Additionally, women have been shown to have greater memory retention of the physical characteristics and appearance of others (Hall, Murphy, and Schmid Mast 2006; Schmid Mast and Hall 2006). This supports the evolutionary theory that greater sensitivity to appearance was an adaptive trait for women because physically more attractive women were able to acquire more desirable males, therefore showing a greater ability to support and protect offspring (Cashdan 1998).

HONESTY VERSUS DECEPTION: "THE EYES ARE THE WINDOW TO THE SOUL"

Recall in our previous discussion addressing philosophy and emotions in chapter 1 that they were viewed more as a distraction to reason and virtuous behavior (i.e., stoicism). Aristotle countered this prevailing view by arguing that emotions (or passions) were virtues of the self and more generally expressions of human nature. Behaviors were a function of physical expressions in our environment that also were influenced by our thoughts and feelings.

More recent empirical work in social psychology and evolutionary psychology does suggest that humans do tend to change their styles of interaction and communication when deception is involved (DePaulo and colleagues 2003; Ekman and O'Sullivan 1991). The idea that gender differences exist with respect to accurately identifying lying and deceit is a popular and controversial topic among both psychologists and anthropologists today. Evolutionary theorists have traditionally argued that distinct skills and traits that have historically remained gender specific (i.e., males with eye-hand coordination; females with collaborative communication skills) and have served important functions throughout human evolution. The idea that women have retained unique skills in their ability to distinguish honesty versus deception has not been supported, however, in the empirical research (Ekman and O'Sullivan 1991).

There have been several indicators that have been associated with identifying deception relative to facial expressions. However, DePaulo and colleagues (2003) completed a meta-analysis that explored deception among several related factors: thinking and feeling cues, detecting false negatives (i.e., liars who are judged to be truthful) and detecting false positives (truthful person who is judged to be lying), facial expressions and nonverbal communication (i.e., body language, such as facial expression, eye contact and body position).

In general, DePaulo and colleagues identified five factors that were positively correlated with deception:

- Actors who attempted to deceive others were noted consistently to avoid direct eye contact when lying ($d = -0.15$);
- The voice pitch of liars was consistently noted to be significantly higher ($d = -0.02$) in pitch than with participants involved in no deception;
- Liars tended to describe and elaborate less details of their fabrication than those who were not involved in deception;
- Liars tended to make more exaggerated facial expressions (i.e., pressed lips) than truth-tellers;
- Verbal descriptions among liars were more inconsistent and illogical than truth-tellers (d = –0.25); and

Liars tended to engage in a sequence of facial expressions and gestures that are described by Zivin (1982) as "plus face" traits: Raised chin (as if an indicator of defiance), raised brows (i.e., "What do you mean you don't believe me?") and directed (i.e., confrontational) eye contact. As research addressing facial expressions indicating emotional states became more prolific in evolutionary psychology, we can assume that some of these skills were used as adaptive tools in identifying others as potentially cooperative and helpful support systems or threats to our own safety.

In sum, the ability to not only detect distinct differences in facial expressions but understand their authenticity within a broad range of human interaction and communication has proven to be an important tool in evolutionary adaptive behaviors. Understanding not only different types of facial expressions but also different types of body language has both played a key role in our ability to understand and manage protective and cooperative relationships with other individuals. Similarly, our ability to detect cues by way of body language and facial expressions as indicators of potential danger, such as aggression, also has had important evolutionary ramifications in understanding how humans have interrelated with each other (Aronoff, Woike, and Hyman 1992).

Those individuals who were able to accurately detect facial expressions and other nonverbal forms of communication had distinct survival advantages over others lacking in this quality. The adage: "Beware of those who bring gifts" should be a reminder that smiles (and gifts) are not always authentic indicators of good will and support. Ultimately, our evolutionary ancestors relied on a consistent and accurate mechanism to identify friends as opposed to enemies. Part of this important function was a result in accurately identifying facial expressions, nonverbal communication and body language (i.e., gestures) to determine whether strangers were friend or foe.

Chapter 6

The Modern Problem of Aggression

All things must change over time as a means of survival. An interesting and early theory of evolution argued that over time all groups with common ancestors (monophylum) tend to actually increase in physical size and stature over time (known later as "Copes Law"). Groups with this common ancestry are referred to as "clades" from the Greek word "*klados*" or literally "same group." This law addressing inherited growth patterns was originally referred to as "Copes Law" named after the American paleontologist Edward Drinker Cope (Rensch 1948; Krieger 2015). As organisms increased in size and stature there were some advantages (i.e., better equipped to fend off predators, more successful reproduction) but also some inherent disadvantages (more dependent on natural resources and a significantly longer offspring dependency on protection from potential threats and predators). The crux to survival, however, was genetic variation and the capacity to evolve and change over time.

Clearly, Darwin's theory of evolution, genetic variation and natural selection supports this premise. Behaviors that may have been highly adaptive in previous millennia (i.e., territorial aggression) may no longer remain adaptive for a variety of reasons and may even become counterproductive to survival. Even the earliest philosophers observed the tremendous potential for human destruction caused by individuals who were prone to engage in aggressive behaviors. Socrates noted the destructive nature of aggression and the vast potential problem that unregulated emotions can create within society and personal and relationships. All behavior should first be examined (via introspection) and analyzed as being ethical and virtuous in relation to public life. Additionally, Socrates argued that all behaviors should serve as a moral compass in one's life and how a "search for truth" is necessary in bringing about individual happiness (Walsh, Teo, and Baydala 2014). The emotion aggression would also be considered problematic by most people because anger and aggression (at least for many people) often interferes or prohibits rational and intelligent thinking processes from occurring in a variety of social and professional situations. Similarly, Plato argued that only a few gifted scholars such as "philosopher kings" (especially not the politicians) in view of their enlightened "reason and wisdom" and search for truth are capable of governing the large masses of people (Plato, *Republic 488*).

The rule of democracy and sense of community is preserved through the efforts of various groups of individuals from different classes (i.e. tripartite) working collaboratively, each possessing distinct skills that are needed, shared and distributed in any society. Plato noted in the *Republic* that if economic gaps between the rich and poor become too pronounced, problems are likely to develop that among the differing classes that will ultimately result in tension, imbalance and even chaos (what we might refer to today as realistic conflict theory). An undisciplined society is vulnerable to exploitation and abuse and typically will result in a tyrant taking command to govern the people.

Norbert Blossner (2007) offers a metaphorical analysis of the three characteristics of the soul offered by Plato (appetite, spirit and reason) with economic levels commonly seen within society: Skilled laborers (i.e., the builders of society, such as masons and contractors), armed or governing forces such as police and army veterans that protect members of society; and the gifted "Philosopher Kings" that make informed choices regarding the welfare of the group itself. The essential task is in the creation of a community that provides opportunities for each group to interact and communicate efficiently with each other for the good of society. Additionally, Plato and Socrates both argued that preserving integrity through wise and informed choices and more importantly in helping create societies that promote individual growth and allow individuals to make their own discoveries that lead to a happier and more productive life.

Even the earliest philosophers (i.e., Plato and Socrates) have argued about the vices of aggression and how human nature is easily corrupted by both power and greed. In Plato's *Republic*, for example, the concept of a true democracy is challenged by the inherent nature of people to control and monopolize resources which ultimately results in a ruling class of tyranny. Aggression and irrational behavior is described by Plato more as an outcome or byproduct of economic inequity and oppression that results in conflict between elite and ruling parties (i.e., a bad tyrant) and the commoners that comprise any given society (Dorter 2006). From this economic inequity other social, democratic and economically related problems develop, such as oppression, discrimination and human rights violations (Schofield 2006).

EVOLUTIONARY DETERMINANTS OF AGGRESSION

Noted social psychologist Carol Tavris describes anger as "the most misunderstood of all emotions" (1989). Anger and aggression are both described as being misunderstood by Tavris in the sense that most individuals perceive anger as having inherently and exclusively negative repercussions in a variety of situations and relationships. Aggression is viewed as a problem because it generally means by definition that someone is capable of intentionally harming someone else (or harming groups of individuals). It also means that people often become irrational and unpredictable if their behavior is controlled primarily out of emotions such as anger. However, if we view aggression from an evolutionary perspective, both aggression *and* altruism appear to share similar qualities in that both are evolved psychological traits designed to protect in-group (and in some cases out-group) members. For example, an individual may use force (aggression) against a perceived attacker in an effort to rescue or assist a genetically related group member (thereby increasing inclusive fitness) or non-related stranger (true altruism). Hamilton (1964) describes this theory in more detail by arguing that altruistic behaviors (cost) are more likely to be selected if the benefit to the recipient (beneficiary) outweighs the risks to the altruist (benefactor): $C < rB$.

Historically aggressive behaviors have served several important functions from an evolutionary and adaptive function. David Buss (2015) has cited several key functions of the human emotion aggression that have clearly played an adaptive role in the survival of humans as an evolving species. Aggression can be instrumental not only in the protection of one's own resources (food, water and weapons) and kin; it can also serve to protect future fertile potential mates as a means of ensuring reproductive success. Demonstrating physical prowess, strength and bravery against competitors not only protects vital resources in one's own clan, but also serves as an effective strategy in identifying status (i.e., power and hierarchy) within one's group and preventing mate poaching which would severely limit reproductive success.

While aggression typically is viewed as a highly destructive action, it can also be carried out as a means to protect and safeguard others from harm. Webster (2008) describes this dual relationship between aggression and altruism as the Kinship, Acceptance, and Rejection Model of Altruism and Aggression or KARMAA. At least from the evolutionary viewpoint, without aggression humans would have been vulnerable in many ways and ultimately less able to survive. Without aggression, for example, humans would have been limited to defending themselves or their kin. Despite the negative implications of aggression, without it humans simply could not have survived given the enormous number of predators that existed during their evolutionary history.

In a recent description of aggression and evolutionary theory, Liddle, Shackleford, and Weeks-Shackleford (2012) argue that a common theme among both humans and nonhuman animals involving aggression addresses sexual selection and parental investment. The authors point out that in virtually every known animal population, some form of aggression is evident. A contributing factor to the common theme of aggression is access to resources (i.e., food) as well as (intrasex) competition among males for fertile/childbearing females. In addition to acts of individual violence and aggression, Shackleford and colleagues note group (coalitional) violence is common among animals as a secure way of obtaining food, especially among carnivores such as lions and African wild dogs (Creel and Creel 2002) and lower primates, such as chimps (Goodall 1979). Thus, aggression distributed among groups of individuals sharing a common goal (i.e., instrumental aggression) has been observed among several nonhuman animal groups and has been determined to be successful in securing vital resources.

The ultimate act of aggression, infanticide, has also been noted in several nonhuman animal groups among both males and females (Janson and van Schaik 2000), and both cases have been noted to support evolutionary theory in the distribution of vital resources. Males committing infanticide may be doing so to improve their reproductive success and future mating opportunities, whereas females who engage in infanticide may be doing so to as a means of protecting their more robust surviving offspring from exposure to scarce resources (i.e., lactation). Liddle, Shackleford, and Weeks-Shackleford's (2012) main argument is quite simple: Violence and aggression (individual, group and war-related conflicts) are common among all animal groups and therefore must serve a variety of adaptive purposes. From a reproductive perspective, violence among males is a relatively common occurrence, where 87% of homicides are committed by males (Lester 1991). Status, access to females and altercations involving reputation are common themes of aggression involving males (Buss 2005) and thus may reduce reproductive success for males when women have shown preferences for males with higher incomes and status. Additionally, altercations among males involving insults and degradation of one's character can quickly escalate into aggression and even homicide primarily as a threat to reproductive access to females, both theories supported in evolutionary psychology (Buss 2005).

MODERN INFLUENCES OF AGGRESSION: MEDIA

The reasons (and explanations) why people have engaged in aggressive behaviors have intrigued social psychologists and evolutionary psychologists for years. Indeed, in some instances altruistic actions to within group members have required an aggressive behavior inflicted to out-group members (Webster 2008). Although at first glance altruism and aggression may appear to be diametrically opposed behaviors with very little similarity, they actually have much in common, as helping one group often requires some form of conflict to another group (Webster 2003). Aggression and antisocial behaviors have remained some of the most controversial topics in psychology, and it seems every few years new theories are being offered to describe the causal factors that have been associated with aggression. As recently as December 2012, some researchers have adamantly denied the relationship between exposure to gratuitously violent video games and violent behaviors (Ferguson, 2012: "*Sandy Hook Shooting: Video Games Blamed, Again*"). Chris Ferguson argues in several recent publications (Time magazine, December 20, 2012) that "there is no good evidence those video games or other media contributes, even in a small way, to mass homicides or any other violence among youth."

Unfortunately, there is a significant amount of *irrefutable* empirical evidence that directly contradicts Ferguson's comments. Numerous studies have in fact documented and supported the fact that exposure to gratuitous violence does in fact contribute to aggression among populations of individuals (Anderson, Berkowitz, Donnerstein, Huesmann, Johnson, and Lintz 2003; Paik and Comstock 1994; Huesmann 2010; Wood, Wong, and Cachere 1991). Additionally, the findings from numerous studies documenting the effects of media violence on youth have even been presented as testimony to United States Senate hearings (Anderson, Carnagey, Flanagan, Benjamin, Eubanks, and Valentine 2004). Exposure to gratuitously violent media has been just one theory that explains increasing violence in society. Additionally, in the most recent study yet available that explores the role of media and aggressive behaviors, the University of Washington researchers have determined that media does influence children's behaviors (http://www.latimes.com/health/boostershots/la-heb-television-intervention-20130218,0,3260824.story).

Conversely, children display more cooperative behaviors when they are exposed to positive forms of media, such as prosocial programming that emphasizes the benefits of group work and collaboration (Greitemeyer, Osswald, and Brauer 2010). The study also found that when children are exposed to programs that emphasize positive and constructive behaviors (i.e., sharing with peers and cooperative behaviors), then the rate of peer interactions characterized by cooperative behaviors also increased. We shall now briefly review several theories that combine more of an instinctual and evolutionary approach in our efforts to better understand some of the more viable explanations of modern aggression. Media and video games are now being used more frequently as important tools to encourage healthier behaviors, such as increased exercise and improved prenatal care practices (Star Tribune article, March 3, 2012: "Healthier living via video games: Click here . . .").

Some of the earlier theorists that have been briefly discussed earlier in this monograph have argued that aggression and antisocial behaviors are in fact *unavoidable* behaviors as they served important evolutionary roles (i.e., instincts) throughout human development. Thomas Hobbes argued in *Leviathon* that without the existence of some form of civilized law (i.e., a social contract), the ability for humans to survive with each other would be futile (Hobbes 1651). Furthermore, as a means of self-preservation, evolutionary history favored a male's highly aggressive actions as a means of improving their ability to reproduce with a significantly higher number of available females. Freud would most likely agree with Hobbes' primitive egocentric views of human nature. Both argue that because humans are essentially opportunistic and irrational creatures, some form of municipal governance is necessary to maintain order in society. This democratically achieved governance would essentially prevent humans from annihilating each other and allow for some semblance of civility and governance.

Where Freud differed with Hobbes was the actually source of the instinctual drives. Freud argued that the essence of the human condition and the primary motivating factors of *all* behaviors was that of sex (i.e., reproductive fitness) and aggression (i.e., protection of resources for kin). All other human behaviors can be reduced to our capacity to reproduce (i.e. instill life) and yet also to take it away through highly aggressive and destructive means. Ultimately the motivating drive of behavior explained in evolutionary terms would be reproductive fitness. If the primary driving force of behavior was to reproduce, this practice would ensure the survival of a particular species.

Freud was a reductionist theorists who emphasized those unconscious forces between the id (pleasure principle) and the superego (moral conscience) would produce a form of tension that only the ego (executive portion of the personality) could resolve. In severe cases where the ego was incapable of resolving extreme anxiety or irrational impulses, then Freud argued that violent and psychotic behaviors typically were the result (i.e., paranoia). Defense mechanisms (i.e., rationalization, projection) according to Freud, were a direct result of the ego's inability to accommodate the demands of the id and the superego.

Indeed, Freud offered the term "*Thanatos*" as the "death instinct" that described a universal drive toward the ultimate balance or quiescence (death). In one of Freud's more macabre quotations toward the later years of his life, he described futility of any person to engage in any type of altruistic or empathic behavior, and he essentially described futility of our own existence in finding meaning in our lives as: "The aim of all life . . . is death" (Freud 1930). The unconscious conflict between the id impulse and the superego often creates anxiety that the ego (i.e., the "executive" portion of the personality) would need to resolve. Freud further maintained that in some cases of extreme anxiety, the ego would become overwhelmed and resort to what is now referred to as "defense mechanisms" of the personality.

While traditional Freudian thinking is considered by most psychological theorists and historians today as obsolete, we briefly mention Freud's theory here as offering some support to the evolutionary theory of aggression as being both a drive reduction theory and instinctual theory. Much of Freud's work is actually based on unconscious mechanisms and the ability for our rational component of the personality (i.e., ego) to satisfy the demands of the mutually exclusive needs of the id (pleasure seeking and animalistic drives) with the moral conscience of the superego. Freud describes all human behavior as an attempt to adapt to the laws that govern society. Without rules and laws that monitor our behaviors, Freud argues that anarchy and civil unrest would ultimately develop and define our own existence. Traditional psychoanalytic theory argues through the process of catharsis that humans are capable of reducing aggressive biological and instinctual drives via exposure to violent media and literature. Interestingly, the traditional psychoanalytic views regarding causal factors of aggression and more currently observational learning theories are in direct contradiction of each other. Observational learning theories argue that exposure to violent media results in increased aggressive behaviors whereas psychoanalytic theories argue the opposite, the exposure to violent movies and literature can actually reduce aggressive impulses via a concept that Freud referred to as *catharsis:* The process of working through potentially violent impulses through socially accepted means.

The term aggression itself requires a more detailed description as we explain various theories that influence how aggression is manifested within society. For our purposes, we can define aggression as any attempt to intentionally inflict harm on another person. Note that the distinction of "intention" is critical here. The frequency in which people intentionally inflict harm on others today is staggering. Current estimates that describe the frequency in which children are exposed to gratuitous violence are alarming. According to recent estimates, there are over 812 acts of violence displayed on televisions per hour, with the average child witnessing over 16,000 murders displayed on television by the time he or she reaches the age of 18 (American Academy of Child and Adolescent Psychiatry 2010).

The evolutionary theory in psychology argues that males often competed for reproductive-bearing females, and males who were more able and equipped to impregnate more females increased their kinship and overall reproductive fitness. Therefore, aggression was a critical and highly useful means of improving our ability to survive and pass on our genes to future generations (Buss and Shackelford, 1997b). Given the fact that males needed to collaborate and reproduce with females (sometimes voluntarily and sometimes by force), it is reasonable to conclude that today there are significantly fewer actions of aggression and hostility among males to females rather than male to male aggression (Hilton, Harris, and Rice 2000). Males are also significantly more likely to aggress against other males typically as a result of competition for finding fertile (childbearing) females and consequently increased reproductive fitness (Fischer and Rodriguez Mosquera 2001) as well as highly impulsive behaviors (Cross, Copping, and Campbell 2012).

The factors that have contributed to aggression throughout human history are highly complex; often many social, economic, and political views coincide with each other to provide explanations why some individuals are considered to be more aggressive than others. Four of the most prevalent and widely used arguments in the descriptions of aggression and their relation to the evolutionary theory are now described: Drive (Frustration Aggression) Theory; Realistic Conflict Theory; Observational Learning Theory (i.e., media); Cultures of Honor, Sexual Jealousy & the Alpha Male.

- Drive (Frustration Aggression) Theory. The classic frustration-aggression theory dates back to the 1939 classic study involving Dollard and colleagues. Dollard and colleagues first worked with animals in laboratories and discovered that as personal space decreases there exists a significant trend or likelihood for animals to aggress. In some situations, food was deprived, and in other situations personal space was limited. In all situations involving space and crowding there was a significant likelihood for the rats to attack each other due to limited space or resources. The classic 1939 research has identified the relationship between environments that prevent individuals from achieving specific goals (i.e., career, educational or even interpersonal) with an increased likelihood to engage in aggressive behavior;

These researchers explored under what conditions was aggression most likely to develop, and they found the universal trait that when humans are consistently deprived of (i.e., provoked) of receiving basic goals, there was a significant tendency for people (especially males) to respond with aggression (Dollard, Doob, Miller, Mowerer, and Sears 1939). Dollard and colleagues initially described the frustration-aggression model as a drive state. Humans are biologically driven to maintain a type of homeostatic balance among our psychological and physiological systems. When we are thirsty, we want to quench our thirst, when we are bored in our environment we typically seek some form of stimulation, etc. Sexual jealousy and reproductive fitness have both played an important role in understanding how the trait aggression evolved among modern humans. Males are significantly more likely to aggress against other males in competitive relationships for available fertile females. Similarly, sexual jealousy and aggression have accounted for the majority of homicides involving males competing for females (Daly and Wilson 1988). Thus, the capacity of aggression and sexual jealousy in securing females for increased reproductive fitness was an important, if not necessary, trait during human evolution that explains conflicts involving what is commonly referred to as "love triangles."

Frustration is not exclusively a precursor of aggression and antisocial behaviors. People may engage in what has been described as a form of "instrumental aggression" where an aggressive act is committed to achieve a particular goal, without necessarily experiencing any anger in the display of aggression. As several researchers have noted, intentional acts of aggression are typically followed by some form of provocation. This form of provocation most commonly occurs when humans cannot achieve or acquire some type of primary resource necessary for their survival or if the provocation is perceived as intentional (Ohbuchi and Kambara 1985).

Throughout our evolutionary history, when our safety and welfare has been threatened, one of the most common (and successful) responses has been to engage in aggression as a means of protecting our resources. Goals that are routinely blocked have also been associated with a higher incidence of antisocial or aggressive actions (Dollard and colleagues 1939).

In addition to frustration, when people feel that they have been treated unfairly or if they feel that they have been threatened by others, they are more likely to act out aggressively. Take for example the employee who has applied for a promotion but has been repeatedly denied or passed over. The employee may feel that he or she is a victim of discrimination and that other employees are treated more favorable for arbitrary reasons. The fact that: a) the employee's promotion request has been denied; and b) he or she has perceived to have been treated unfairly are two combining factors that often have been associated with workplace violence (Folger and Baron 1996).

Frustration-aggression theory has remained one of the most popular and widely accepted theory of aggression. In terms of survival, it was critical that humans retained the skills and capacities to identify and retain valuable resources, such as food, fertile mating partners, and safe living conditions. Aggression played an important role in protecting these assets. An initial emotional response, such as frustration, that was followed by anger was a direct mechanism that evolved as a means of protecting ourselves from harm. Aggression clearly holds an adaptive value in achieving goals that are not readily available to us, and the development of frustration as a precursor (or dependent variable) to aggression may have proven to be a highly adaptive trait relative to our survival and reproductive fitness.

- The Realistic Conflict Theory. Employment, housing, and access to vital resources are all key indicators not only of our material success, but more

> importantly indicators of our ability to provide the necessary resources to enhance development of our family and kin. When these key resources have been systematically blocked from us unfairly, or if we perceive that we simply cannot achieve these resources, we are more likely to respond aggressively in specific types of environments, such as the work place or personal relationships (i.e., domestic violence or spousal abuse) if we perceive our spouse or significant other has been unfaithful to us (Folger and Baron 1996).

The realistic conflict theory also argues that a common factor that is associated with aggression and conflict among different ethnic groups is the dispersal of available important commodities, such as employment, housing and even basic living necessities such as food and water (Bobo 1983; Sears and Kinder 1985). The strategic use of aggression was an effective means of stockpiling and securing resources for one's own group. The relationship between frustration-aggression theory and realistic conflict theory is relative to evolutionary psychology because each theory addresses negative outcomes with one group (i.e., the dominant group) not achieving specific goals that are perceived to be inherent in the goals of ideal living standards. Realistic conflict theory argues further that typically marginalized groups or minority groups are identified as the primary threat to access future necessary resources and are subsequently are more likely to become victims of out-group bias and prejudice (Miller, Maner, and Becker 2010).

The Realistic Conflict Theory also describes why underrepresented groups are often more vulnerable to being victimized and exploited through competition in the development of scarce resources, such as employment, housing and even the basic necessities to sustain life, such as food. Research has documented on numerous occasions that marginalized groups are more likely to be victimized in times of economic recessions and crises (Kinder and Sears 1981; Hovland and Sears 1940). More recently, hate crimes have been defined as "any crime in which the victim is selected due to actual or perceived race, color, religion, disability or sexual origin" (Sullaway 2004).

As Park, Shaller and Van Vugt (2008) have stated that a better understanding of our own evolutionary history can provide insight to group dynamics and behaviors. Throughout millions of years of evolutionary history, certain behaviors among group members were favored as adaptive as they helped to prevent harm to the group itself. Strangers who showed little phenotypical resemblance to us may have posed significant threats to our safety. Current persistent social and psychological problems such as ethnic conflict, ethnocentric ideology, prejudice and discriminatory behaviors may have their roots in our evolutionary history primarily as a protective mechanism in maintaining contact with those who are more likely to phenotypically resemble us.

Furthermore, an improved understanding of evolutionary psychology may help us to better understand group dynamics in a variety of ways (Park, Shaller, and Van Vugt 2008; Van Vugt and Schaller 2008). The better we understand our own evolutionary history, the better we are able to understand the positive (and negative) dynamics of group process and interaction. For example, a better understanding of group dynamics may help us to not only reduce systemic race-related problems such as ethnic conflicts and hate crimes, but may also help us to shape more prosocial behaviors that existed during our own evolutionary history such as altruism. The authors also make important contributions that help us to understand the relationship between evolutionary psychology and key behaviors that can help to improve the relationships among individuals from different economic, religious and ethnic backgrounds.

Often an aggressive action may be displaced (i.e., engaging in an aggressive action to someone or something not directly responsible for the transgression) or direct (engaging in an aggressive action with the person who is perceived to be the original provoker). In the realistic conflict theory, a type or form of aggression (i.e., hate crime) is directed toward a specific member of an underrepresented or marginalized group. Realistic conflict theory is described as: "a type of prejudice typically involving conflict or aggression due to direct competition over scarce resources" (Baron, Branscombe, and Byrne 2008). An economic recession often results in fewer jobs and overall increases the cost of living. Unfortunately, as competition over few jobs increases, the likelihood of conflict and aggression (based on ethnicity) is significantly increased. Unfortunately, numerous empirical research has shown an inverse correlation between increased hate crimes and reductions of employment that are typically associated with economic recessions (Hovland and Sears 1940; Mintz 1946).

- Observational Learning Theory, Media and Evolutionary Adaptive Mechanisms

There have been many theories describing aggression from several perspectives, and the two theories consistently surface as contributors to aggression: Aggression as an action that is vicariously influenced (i.e., "learned") from environmental influences (i.e., role models) and aggression as an inevitable and biological outcome of instinctual drives and human nature. Both perspectives may be explained from an evolutionary adaptive perspective. Some of the earliest researches available addressing the etiological factors and aggression suggest that aggression was an important emotion that helped individuals to survive when exposed to a variety of threats as well as securing vital resources. An important skill that clearly aided our ability to survive as a species was in our ability to visually emulate and imitate those behaviors that we perceived to be both desirable and important to our well-being and survival (Hayes, Rincover, and Volosin 1980).

In the former example, humans have adapted to various types of environments in their unique ability to emulate and copy behaviors that they have observed from role models. Primates, for example, may teach their offspring how to escape from potential predators by first scanning the environment before searching for food. In the event that a predator does attack, the primate can escape by quickly climbing a tree. The offspring has learned that first scanning is essential to detect potential predators and then always maintain an avenue of escape by maintaining proximity with a tree. Without developing some rapid form of imitation of adaptive behaviors, organisms would have quickly become vulnerable to a broad range of predators. Emulating adaptive behaviors of role models (i.e., family or kin behaviors) through observation was a vital behavior to help ward off threats and predators.

The tendency to imitate those behaviors from perceived positive role models has remained with humans throughout our evolutionary history. Humans are not unique in their ability to imitate what they see in their environment. Various primates, such as the neonatal Rhesus Macaques have been observed to imitate not only adult Rhesus Macaques, but also humans in various facial expressions as well as behaviors (i.e., lip smacking and gestures).

Learning by observation has remained one of the most studied topics in psychology and evolutionary theory. The ability to not only observe models but develop the capacity to imitate their behaviors was a key factor in survival and adaptation. Behaviors can be learned vicariously (i.e., by watching others) and demonstrated in positive ways or negative ways. It should also be noted that learning by observation can explain various styles of behaviors, both gratuitously violent behavior and positive styles of interaction, such as cooperation and sharing of resources.

An additional theory to exposure to media violence is the fact that the more individuals are exposed to aggressive-content material, the greater the likelihood of desensitization. When individuals are exposed to larger amounts of aggressive-content media, there tends to be an increase in the belief systems of the justifications of violent behaviors due to previous exposure.

Recent estimates of exposure to gratuitous violence (i.e., murders) among children living in the United States are staggering. By the time a child just finishes primary grade school (typically age 12), the average child has seen over 8000 murders on television and over 200,000 acts of violence by the time he or she reaches the age of 18 (http://www.csun.edu/science/health/docs/tv&health.html). The recent events involving aggressive behaviors as portrayed by the media (i.e., news events) also have increased significantly in the United States (Huesmann 2007). Recent scientific data conducted in several large studies confirm the hypothesis that exposure to violent programs not only increases aggressive behaviors, but can also significantly increase thoughts that may provoke violent behaviors (Kirsh 2012). As technology has steadily increased, so has access to violent programming and recreational video games among younger populations. The content, information programs that are now available to younger audiences is more graphic, stimulating and interactive than previous electronic entertainment (Donnerstein 2011).

As media continues to portray and associate positive role models with a broad range of violent crimes, it is not surprising that overall acts of violence have steadily increased over the years, with over 590,000 homicides occurring in the United States in 2007, *averaging 161 deaths per day* (US Census Bureau, Statistical Abstract of the United States, 2012). Clearly, our evolutionary practice of learning by observation has influenced our capacity to engage not only in less desirable behaviors such as violence, but more positive, prosocial and cooperative behaviors as well. If we wish to truly reduce the numbers of violent acts within our society, we need to closely monitor who we portray as positive role models within our society. If we as a scientific community can encourage the entertainment industry to begin showing models in a more positive (i.e., less violent) image, according to the basic rules of observational theory we should see significant reductions in violent behavior.

Unfortunately, the effects of media violence as described through the observational theory of learning have both short-term and long-term effects on the individual who is exposed to it. Recent research has shown that as the brain is developing in early childhood, neurons become associated with specific forms of images that the child may be exposed to. Neural priming effects develop when children are exposed to novel or new stimuli and the process is described as spreading activation. Spreading activation occurs when neurons become associated with specific types of stimuli (i.e., violent media). These neural priming effects tend to become more developed over time and become associated with each other. When a child is exposed to a violent image in the media, neural images and nodes develop that increase the likelihood that violent behavior is more likely to occur (Huesmann 2007). Despite over 40 years of empirical research that clearly documents the role of media violence and aggressive behaviors, there still are those who insist that no relationship exists between exposure to gratuitously violent materials and aggressive behaviors.

Perhaps one of the most well-known and documented studies that empirically validates the role of observational learning, media and human behaviors is that of Albert Bandura and the now famous "Bobo" Doll experiment. The 1960s era was both one of positive forms of change relative to civil rights and violence. Many researchers in psychology explored the question of what are the causal factors of aggression. Was Freud correct in his theory stating that individuals are essentially born with a destructive instinct (i.e., Thanatos), or does the environment in some way contribute to violent behavior? In 1961, Albert Bandura began a campaign of experimental research that explored this very question. He argued that the kinds of things that children are exposed to can in fact contribute to violent behaviors. Current research consistently shows that when children are exposed to models who are rewarded for cooperating with others, sharing, and demonstrating prosocial behaviors, these behaviors are imitated (Bandura 1977). Previous research addressing the capacity of children's ability to share, cooperate and reason had been conducted by Jean Piaget (1932), but Piaget argued that children were limited to stage and sequential forms of development that can only occur over periods of time (i.e., Preoperational stage and Concrete Operations stages).

Bandura argued in several studies that social learning theory allows for children from a broad range of ages (from ages 4 to adolescence) to watch models (for as little as 20 minutes) engaged in specific desired behaviors, and children would imitate these behaviors. Conversely, in a series of experiments where the adult role model did not hit or attack the Bobo doll, the children typically did not hit the doll and imitate the adult role model.

In vicarious reinforcement, children watch models engaging in cooperative behaviors and are rewarded for their behaviors. This form of reinforcement has been shown to hold powerful influences in correcting children's behaviors. Bandura determined that when children are exposed to positive role models, they are significantly more likely to engage in cooperative and supportive behaviors, as well as display more complex behaviors involving empathy and prosocial concern for others. Unfortunately, the reverse has also been shown to be true—where negative role models displaying egocentric or aggressive behaviors are just as easily imitated (Hastings, Karas, Winsler, Way, Madigan, and Tyler 2009). Interestingly, the evolutionary trend to imitate behaviors is not limited to visual stimuli. Recent research also indicates that violent lyrics in music have also been shown to be correlated to significant increases in violent behaviors (Fischer and Greitemeyer 2006).

Lower primates and humans have imitated behaviors throughout evolutionary history. The capacity to imitate by way of observation clearly holds an adaptive value as a resource to learn within a variety of different types of environments. More recent research (Heyes 2012) argues that the fundamental characteristics of social learning can also occur in asocial learning among primates within three domains: Stimulus enhancement, observational conditioning and observational learning. In earlier research, Macaque monkeys were shown a stimulus (i.e., food) that was manipulated and hidden in a different location under laboratory conditions. The macaques followed (and successfully identified) the manipulated object and ignored the other object that was not associated with food. In observational conditioning, monkeys that have seen other monkeys avoid a feared stimulus (i.e., a snake) will also mimic that behavior and avoid snakes. Observation of one's kin will be imitated and is considered highly adaptive if the feared stimulus poses a significant threat to the safety of other same-species animals (Cook, Minka, Wolkenstein, and Laitsche 1985). The primates (i.e., rhesus monkeys, Macaques, orangutans, and gorillas) are examples of some primates that have imitated a broad range of behaviors in their natural (and sometimes unnatural) habitats. Finally, in observational learning, observation of one specific type of behavior that is perceived to be adaptive and helpful has often been emulated by other animals. This is especially true among goal-directed behaviors where an animal sees another animal's behavior as being somehow useful and fortuitous to its own safety:

- Cultures of Honor, Sexual Aggression and the Alpha Male: "*Lavar a Honra com Sangue*" ("Wash the Honor with the Blood")

Males have evolved to become significantly more aggressive than females and males are also most likely to be the victims of their own violence (Daly and Wilson 1988). Fewer competing males in any given group provides greater access to fertile females which ultimately increases reproductive fitness for those surviving males. Exposure to violence and aggression increases the likelihood of repeated episodes of similar violence, and cultures can increase the likelihood of aggressive behaviors among younger male populations. This phenomenon has been referred to as the "young male syndrome" (Wilson and Daly 1985).

The topic of culture and aggression has recently become popular topics of study and research (Vandello and Cohen 2003). Specific types of aggression seem to be directly correlated with more traditional (i.e., "machismo") cultures, especially when the topics of fidelity, honor and trustworthiness are addressed. In many of the more traditional cultures (i.e., Latino and Southern cultures) that maintain gender-traditional norms, a significantly greater likelihood of aggression is more likely to develop especially in the cases of infidelity, honor and trust. The term "machismo" should be addressed here in relation to the "Alpha Male" as both concepts have recently become more prone to negative interpretations by the general public.

The term machismo generally describes a type of behavior among men of Mexican culture. It usually involves (but is not limited to) womanizing, forcefulness, sexism and negative characterizations of masculine behavior. Arciniega, Anderson, Tovar-Blank, and Tracey (2008) have described a more positive version of the term machismo as "caballerismo." Caballerismo is described as a positive form of machismo that describes the male (not necessarily Mexican or Hispanic) as protective, nurturing, patriarchal and generally more positive in established familial relationships. Similarly, the Alpha Male is described more in popular literature as a male leader who exercises dominance and control over other males and females in some form of a group situation (http://www.wisegeek.org/what-is-an-alpha-male.htm).

The Alpha Male typically displays superior skills and abilities in the development of necessities, such as food (in evolution) and income and providing a high standard of living (modern males). The Alpha Male was vital in the survival of clan members during evolutionary history. The Alpha Male protected clan members, directed migration and living patterns and resolved internal squabbles and conflicts within the clan itself. Essentially the Alpha Males determined the lifespan and viability of the members of the clan itself.

Previous theories describe aggression as primarily an event that is essentially influenced through a combination of environmental and personality factors. Evolutionary theory historically has favored theories that have described aggression from a more utilitarian perspective. Individuals (typically physically stronger males, such as the Alpha Male) that perceived their own kin's safety being threatened or challenged would respond in a physically aggressive manner as a means to ward off (or in some cases eliminate) the perceived threat. Responding to perceived threats with aggression and physical harm clearly held several benefits throughout human evolution.

Current (and past) research consistently has shown a significantly stronger likelihood for males to become more engaged within a variety of different types of aggression primarily as an adaptive resource from our evolutionary background (Archer 2004). Interestingly, research also shows some gender differences not in the display of aggression, but rather how the aggressive action is manifested (Eagly 1987). While males are more likely to engage in direct aggression, women have been shown to display indirect or "instrumental" forms of aggression (Eagly and Steffen 1986). The current description of aggression attempts to explain how aggression may develop primarily from cultural factors where violence may be rewarded or reinforced in light of specific cultural taboos (i.e., infidelity and cuckolding).

Cultures of honor have also been shown to be instrumental in the development of hyper-masculine traits, and these characteristics are considered essential in modern evolutionary theory (Hill and Hurtado 1996). Physical contests that displayed male strength, virility, and bravery were common throughout evolutionary history and played important roles in how females were both selected and rewarded for greater reproductive fitness. This fact is further supported through a phenomenon that Buss (2008) refers to as "reproductive competition"—a phenomenon that is demonstrated among several species including primates.

Historically, males have competed in a variety of forms of competition that have proven to be a key incentive in female receptivity. The Alpha male showing the greatest degree of stamina, physical strength, and perseverance have historically been provided the greatest access to reproductive females, thereby improving future prodigies and reproductive fitness (Buss 2008). This practice is still continued today among males in competitive sporting events and physical competitions. Individualistic cultures that emphasize winning in highly competitive situations are actually providing opportunities for males to distribute their physical strength, and therefore ultimately improve their overall reproductive fitness. The means have changed, but the end result is the same as it was among Neanderthal males.

While direct physical contact among Alpha males (at least in many modern societies) is discouraged, physical competitions still exist today that allow Alpha males to acquire several receptive female mates, such as sporting events and other physical contests. Less physically adaptive Alpha males have also been noted to compete with stronger males in indirect ways. Open criticism and public humiliation are some methods that do not require physical prowess among weaker males but may also make the weaker male more appealing to females (Wilson and Daly 2004).

Darwin's theory of natural selection in explaining the basic principles of human existence and survival proved to be an important component in the development of modern psychological theory. Adaptive features and characteristics that facilitated one's own survival were transmitted to future populations, thereby increasing reproductive fitness of one's own genetic pool. What Darwin did not anticipate, however, was the problem of phenotypical characteristics that, at least at first glance, did not seem to enhance one's chances of survival via natural selection.

The problem was the male peacock (genus *Pavo* of the pheasant family, *Phasianidae*) and its colorful display of feathers that directly contradicted the normal camouflage of colors in their habitat. Darwin agonized over this biological display of colors as it seemingly appeared to be a direct contradiction to the basic principles of natural selection and evolutionary history. How could bright and colorful displays of feathers actually serve as characteristics that aided and facilitated survival, when virtually other birds and reptiles were essentially focusing on camouflaging with the natural environment?

It wasn't until Darwin further investigated the long-term consequences of the colorful display of feathers that allowed him to discover the answer to this evolutionary riddle. Darwin soon came up with a second theory of natural selection, one that he referred to as "sexual selection" (Darwin 1871) as the answer to the bright and colorful plumage of the peacock. The bright colors of the male peacock served to attract females and thus improve the peacock's ability to reproduce and increase survival of offspring. Similar to the "cultures of honor" where males often compete with other males by demonstrating muscular development as a means of attracting females to produce healthy offspring and increase overall reproductive fitness (Schmitt and Buss 1996).

OPERANT CONDITIONING, AGGRESSION AND THE EVOLUTIONARY THEORY

Perhaps one of the most dynamic and controversial of all modern psychologists who have described learning in behavioristic terminology is B. F. Skinner. Skinner is typically credited with discovering the basic principles of operant conditioning. Operant conditioning is defined as a type or form of learning that is based on the consequences of our behaviors. The basic principles are very simple in operant conditioning, as the consequences of our behaviors are the primary determinants of future behaviors. In this way, then, environmental situations have profound capacities in defining our behaviors, depending whether or not the consequences are positive or negative to the subject.

According to Skinner, it is impossible to determine what types of cognitive activities are occurring inside someone's head, and therefore it is only logical and reasonable to assume that behaviors can only be judged and determined by overt behaviors (i.e., the "empty organism" theory). The circumstances that prompt behaviors (or antecedent conditions) as well as the consequences of those behaviors are the primary determinants of future behaviors. If a child observes an adult role model being rewarded for an outwardly aggressive action, then the adult serves as the model and the child will imitate the behavior. This behavior further becomes more likely to reoccur in the future if the child is rewarded for his or her behavior and if the environmental cues trigger the same response.

Skinner did not deny the existence of thoughts or cognitions in daily human functioning. He did contend, however, that these "thoughts" or "ideas" that were contained in the mind were simply impossible to empirically validate, so they should not be studied at all in science. Skinner also notes that behaviors are emitted from the organism as opposed to being elicited. This distinction is fundamentally important in Behaviorism, as emitted behaviors indicate that the organism voluntarily engages in some form of behavior as opposed to the experimenter manipulating or "eliciting" the behaviors (similar to Pavlov's research with dogs—Pavlov elicited the responses by associating the sound of the bell (CS) with the UCS (food)). Skinner often referred to human behaviors as that of "machine like" and predictable. Skinner also commented that humans create machines within their own likeness and image, and that the distinction between automata and humans is rapidly diminishing given increasing technology and production in the modern world.

The world of technology and social media is especially changing at a rapid pace and perhaps more importantly the role of technology is having a radical impact on the quality and level of human interpersonal relationships. Today more than 20% of individuals meet each other online, and many individuals feel that social media is rapidly becoming the favorite means of social interaction and intimacy. Skinner referred to humans as "puppets" that are controlled by environmental stimuli, and would object to the notion of free will or even the capacity or possibility of humans to control their futures as environmental consequences are already shaping our destiny.

EVOLUTIONARY SUPPORT FOR SKINNER'S WALDEN II

Despite B. F. Skinner's highly reductionistic and mechanistic views regarding human behaviors, his work does show significant human capacity for positive growth and change. The key to human growth and change, according to Skinner, has nothing to do with internal perseverance, temperance or individual "will power." The key to adaptation and change lies within the environment that humans live and work in. Environmental conditions, antecedent and consequential factors primarily determine the likelihood of behaviors repeating themselves or becoming extinguished. Both evolutionary psychology and Behaviorism focus exclusively on the relationship between the environment and human behaviors. Evolutionary psychologists argue that modern human behaviors (both prosocial and cooperative) and traits evolved primarily from an interaction between the demands of the environment and how groups used those resources to their advantage. Those groups capable of engaging in collaborative and cooperative behavior significantly increased their reproductive fitness and thus were subsequently better equipped to survive and pass on these traits and characteristics to future generations.

Similarly, the Behaviorists argued that consequences of our behaviors within the environment in which we were exposed to had important ramifications in shaping future behaviors. For example, those smaller groups of individuals that learned the advantages of cooperation while working and hunting were quickly rewarded with life-sustaining food—a commodity that clearly brought individuals together when survival conditions were most challenging. Thus, both Behaviorism and Evolutionary Psychology emphasize the specific characteristics of the environments as critical factors in shaping human behaviors and attitudes toward cooperation as well as human phenotypical characteristics (i.e., physical appearance). The Behaviorists would have little in common with some of the classic Greek philosophers (i.e., Socrates and

Plato), as these philosophers discussed primarily internal (i.e., pre-existing) traits and qualities that defied empirical observation and analysis.

Choose your specific environment and human behaviors will change based on the antecedent and behavioral consequences that are associated with that particular environment. Perhaps Skinner's views and theories in psychology are not as dramatically appealing as other theories focusing on changing human behaviors because Skinner argues that most behaviors can change if we select the right types of reinforcements. For example, an individual who may be trying to stop drinking alcohol should not socialize with peers and friends in establishments that serve those kinds of beverages, or someone trying to lose weight should not go out with friends for lunch but rather select a different type of environment so to socialize in (i.e., a gymnasium, going outdoors for a walk, etc.).

EVOLUTION, COOPERATION AND REDUCED AGGRESSION

In conclusion, aggression remains a highly complex behavioral trait that clearly has served an adaptive need during human evolution. While aggression was an adaptive trait during human evolution, today aggression remains problematic in resolving conflicts that are inevitable in society. However, there are several highly effective and simple strategies that can be implemented within society to reduce rates of aggression. For example, communities and workplace organizations that are developed and structured in a way that encourages individuals to participate in different types of organizational support programs can influence how aggression is displayed in different types of work place settings (Schat and Kelloway 2003). Additionally, how groups within educational and community environments are developed can reduce the likelihood of specific types of violence and aggression (i.e., bullying on school campuses) (Horne, Stoddard, and Bell 2007). Communities that encourage individuals to share their unique skills and participate in community development activities are essentially providing a very basic need of interdependency that has evolved among humans for literally millions of years. Humans have evolved with a very basic need to trust and develop a unique sense of community and connectedness that provides the social glue that connects us with each other.

Chapter 7

Evolutionary Psychology and Gender Differences

Are Men Really From Mars?

Differences in behaviors and equal rights among men and women are certainly not considered new topics in psychology or philosophy. In the classic *The History of Animals* (Book IX, Part I), Aristotle describes significant personality differences in both the temperament and behaviors among men and women, where women show more "compassion" than men but unfortunately are also described as having more personality flaws, such as jealousy, argumentative dispositions and being more "deceptive and mischievous" by nature (Aristotle, 350 BCE). Despite the fact that Athenian females during the late fifth century were deliberately not provided with any means of education and even considered to be "incomplete males" or "property" by the patriarchal ruling hierarchy, Socrates (and especially Plato) had very progressive views of women and the roles that they should hold in society (Nails 2014). For example, Socrates frequently recognized that many of his teachers were indeed women, and noted that he was taught rhetoric from Aspasia from Miletus. Similarly, Plato argued for the need for women to engage in civic roles and suggested in the *Republic* a role for the "Philosopher Queens" as well as "Philosopher Kings."

The term "hysteria" originally was introduced by Hippocrates used to describe women who were prone to excessively emotional and irrational behaviors. The term is derived from the Greek "hystera" (or "wandering uterus") and today is even considered to be a form of sexual disorder described in the American Psychological Association's Diagnostic Statistical Manual IV (revised). Sadly, the bias and discriminatory behaviors that have plagued women since the early Greek era still persist in modern society among many disciplines and employment practices (Unger 2014), and the debate among gender differences continues well into the twenty-first century.

A common and interesting discussion among students is that of gender-related differences in human behavior: Do men and women differ in their behaviors and if so, why? In my courses in psychology students are often asked to imagine two groups of people (one male group and the other group female) and both of these groups of individuals are asked to repair something (i.e., a toaster or a bicycle) that typically involves the use of tools and reading instructions. The next question is the provocative one—would we most likely see differences in the *process* in how each group performed the task? Most of the time the majority of students would clearly say "yes"—males and females do tend to function differently relative to communication, collaboration and group process (Arnon and Ramscar 2012), but the interesting question is in asking *why* and *how* these differences exist.

Many scholars in evolutionary psychology have argued that the reason why males and females often think (and ultimately behave) differently is that they have faced different types of situational and adaptive problems relative to reproductive fitness and survival, and over the millennia these gender-distinct styles of behavior have become genetically "hard wired" into our brain (Buss 2012; Easton, Schipper, and Shackleford 2007). Distinctly different gender-related strategies have been used successfully over millions of years and these unique adaptive problems determined whether or not organisms thrived or perished in the most hostile environments (Barrett and Kurzban 2006).

An equally provocative question addresses gender-related differences in sexual preferences and attraction. What do men and women find attractive (and repulsive) among one another? In this chapter we will discuss the evolutionary components of sexual attraction and how specific sex-related characteristics have evolved over millions of years. There are important reasons relative to our own evolutionary history that explain why we perceive some individuals as particularly attractive and others less attractive, ranging from physical body type and structure (Voracek and Fisher 2006) to facial structure (Henderson and Anglin 2003). Even kissing in romantic relationships has been determined to have an evolutionary component, as kissing may be an ideal method of determining oral health and hygiene (Symons 1995) as well as a technique in identifying compatible mates for reproductive purposes.

In a recent study conducted by Gordon Gallup and David Frederick (2010), the topics sexual attraction and evolutionary theory were addressed focusing more exclusively on the adaptive reasons *why* we find specific body types (i.e., facial structure, shoulder-to-hip ratio, skin texture, etc.) most appealing and attractive. The authors essentially came to one summarizing conclusion: What is perceived to be physically attractive to each of us (cross-culturally) has clear adaptive and reproductive value. Women with specific body types (i.e., what is commonly referred to as the "hour glass" figure) have traditionally been considered to be more attractive. For example, a waist-hip-ratio (WHR) of .7 (narrow waist with broader hips) is considered (and ranked) by males cross-culturally as more physically attractive (Singh 1993) than women with different body types. According to Gallup and Frederick, the reasons why males cross-culturally ranked women with low WHR's probably has more to do with higher successful birthing rates (broader hips generally mean lower infant mortality rates with a larger birth canal), regular ovulation and menstrual cycles and overall robust reproductive health (Van Hoof, Voorhorst, Kaptein, Hirasing, Koppenaal, and Schoemaker 2000). Similarly, males are perceived by females as more sexually attractive with a higher level of shoulder-to-hip ratio (SHR), suggesting a higher degree of musculature for protection of females when incapacitated through pregnancy.

Another interesting topic that is related to sex differences and behaviors is in collaboration and communication. For example, when all men and all women work separately in groups, is the nature and style of their work habits different? Do women (or men) communicate and exchange information more equally with one another when working together on projects or does one individual (i.e., typically the "alpha male") tend to dominate the behavior within the group? If a project requiring interaction and collaboration was given to a group of men and to a group of women, do you think that the different groups based on gender would have a different style of completing the project? Most people would clearly indicate "yes" and have a list of anecdotes and humorous examples supporting their claims.

If gender differences do exist relative to styles of communication and cooperation, are these differences due to an innate and biological drive based on early prehistoric competition for key resources necessary for human survival? Or, are these gender-related differences based on cue from our environment and become essentially learned phenomenon based on our experiences in the outside world? Gender research has traditionally remained a compelling topic of study in several sub-disciplines of psychology (i.e., social psychology, developmental psychology, etc.) and more recently evolutionary psychology. The topics of gender-related behaviors have recently evolved from situational factors (i.e., social learning theory) now to topics exploring the actual origins and etiology of gender-related aspects of behavior.

More specifically, Gabriel and Gardner (1999) have identified gender traits with two aspects of behaviors, including those that separate one from their group (independence) and one that describes how individuals are connected to their group (interdependence). Traditional research addressing gender differences have focused on environmental and socialization as key components that influence and significantly contribute to gender differences in group interaction (Eagly 1987). Specifically, Eagly and colleagues have argued that gender differences in social behaviors have primarily been attributed to status and traditional roles assigned by social norms. "Independent" characteristics are typically described as masculine and emphasize autonomy and distinction from group-related activities (Geis 1993), whereas traits described as "interdependent" have been described as a female-oriented trait emphasizing group needs and how individuals relate to the larger segment of a given population (Markus and Kitayama 1991).

An ongoing controversy in social and evolutionary psychology is *not* whether or not gender differences actually exist (they do—compelling evidence has established this), but rather what are the contributing factors that has promoted gender-related behaviors from developing within a variety of different types of social and developmental situations (Eagly 1987). In a series of experiments testing relationships between gender and interdependence, Gabriel and Gardner (1999) explored the context (i.e., self-construals) in which interdependence was most valued between male and females. Women were significantly more likely to provide self-descriptions that were identified with close (i.e., intimate) relationships, whereas males provided a significantly higher degree of group membership that included a collective aspect of interdependence. Additionally, the findings of Gabriel and Gardner (1999) provide support for evolutionary theory and interdependence actually for *both* genders. Males who possess greater skills in collective orientation and group status would be more likely capable of attracting females that would inherently increase reproductive success. Similarly, females possessing higher capacities of relational skills would have access to greater resources in groups, thereby improving chances of survival for their offspring. Both of these traits (relational skills for women and collective group membership/status for males) would be highly beneficial to both genders in attaining mates for offspring and reproduction (Baumeister and Sommer 1997).

In summary, the important questions that both social psychologists and evolutionary psychologists should be asking are how interdependency influences gender-related behaviors. A dual-approach of examining social roles and status combined with evolutionary adaptive mechanisms can provide valuable insight to these areas of study.

COMPETING REPRODUCTIVE DRIVES: WHAT'S IMPORTANT TO MEN AND WOMEN?

Current evolutionary theorists would argue that the primary reason why sex-related differences continue to exist today is that both men and women have adapted to behaviors that have been best suited toward their own survival and reproductive fitness. Males historically have preferred short-term relationships as opposed to women preferring long-term relationships as a means of protection and child-rearing practices. Males preferred short-term encounters with females as a means of improving reproductive fitness, so finding a variety of receptive (and in some cases non-receptive) fertile females enhanced the likelihood continued kinship in future populations.

David Buss (1989) conducted a meta-analysis that explored sexual attraction within numerous cultures (over 37) where there was striking similarity in what men and women found both sexually attractive and important in relationships. Buss found near-universal support for males who were attracted to not only a specific female body type (i.e., body mass index) relative to height-weight ratio, but also a waist-hip ratio most suitable and supportive for healthier childbearing productivity. Women, on the other hand, were universally attracted to males with a stronger and healthier physical build (i.e., broader shoulders) as well as a preference for their male partners to acquire high earning potential. The near universal link between physical attraction, height-weight ratio and general physical traits and characteristics cannot be summed up as coincidence or chance. Evolutionary theory as Darwin described it accounts for these universal gender preferences for specific reasons—reproduction and survival. Physical traits that were most closely linked to fertility and successful birth rates were perceived as desirable among males, and those physical traits that were linked to physical strength and virility are perceived as desirable among women.

As Charles Darwin would have also indicated, nature selects those traits and characteristics that were most likely to be passed on to future generations, and that these characteristics were not only phenotypical (i.e., physical traits that are observed, such as hair, skin and eye colors) but also genotypical traits (genetic characteristics and DNA) that were most likely inherited in future generations.

An ongoing debate in psychology (as well as several other disciplines) addresses the issue of not only the causal factors of behaviors (i.e., nature vs. nurture), but also the factors that influence how men and women think and behave. As you may recall, for several years during the reign of Behaviorism, people argued that all of human behavior is essentially learned and a consequence of environmental consequences. This would be true when addressing the origins of gender differences. Men, for example, are rewarded for behaving in traditionally masculine way (i.e., aggression) whereas women have been rewarded for those behaviors that reflect a more nurturing, communicative and interdependent nature. The flip side of this argument is that the reason why we see many gender differences in a variety of social and interpersonal situations is not necessarily a result of what has been taught to individuals, but rather that these key behavioral differences proved to be valuable to both sexes in terms of survival and reproduction.

According to the evolutionary perspective in addressing human behaviors, the reason why gender differences exist in the first place is that each sex faced unique challenges and problems relative to their own capacity and ability to survive within the environment in which they lived. Females, for example, were faced with the responsibility of childbearing and child-rearing (i.e., limited mobility as well as limited access to resources during child-rearing activities). Females in search of a mate had many things to consider relative to her (and her offspring's) health and welfare: Her male mate needed to be willing to assist with life-sustaining responsibilities (i.e., acquiring food) as well as possess physical strength to ward off predators and threats to safety. Finally, an ideal male mate would have been willing to remain with his female mate and not abandon his offspring.

Additionally, females also needed to establish bonds with other females within the same clan. Like males, females also had pecking orders or hierarchies. Evolutionary psychologists today will argue that one reason why gender differences among men and women are particularly salient in communication and cooperation is that women needed to be more resourceful than males and they needed communication skills in working with other women. Similarly, establishing a network with other females that was highly interdependent and communicative in securing resources such as food and shelter helped establish significant changes in how man and women interacted with each other and established important community and clan bonding skills.

Evolutionary psychologists would explain the significant gender differences in aggression due to biological (i.e., hormonal) changes as well as how aggression proved to be beneficial to males in terms of securing scarce resources (i.e., food) as well as reproductive fitness (i.e., women during peak childbearing years). So—the question remains—what are the causal factors addressing gender differences in behavior? In a meta-analysis of cooperative behaviors between men and women, Daniel Balliet, Shane Macfarlan, Norman Li and Mark Van Vugt (2011) determined no conclusive evidence to suggest that significant *individual* differences exist regarding the existence of gender-related cooperative behaviors. However, in the same study the authors did conclude that the context of environmental factors did play a significant role in the manifestation of cooperation, and that gender-related differences did exist. For example, where men and women were shown individually to show similar cooperative behaviors (i.e., willingness to help complete a project), groups of men were significantly more likely to cooperate than groups of women (i.e., men working with other men showed the highest degree of cooperation, whereas ironically women working with other women showed the least cooperation).

Additionally, findings by Balliet and colleagues supported the evolutionary theory regarding cooperative behaviors in small groups or clans. For example, males tended to evolve with a specialized ability to hunt, track and harvest game for consumption whereas females evolved with a specialized ability to forage, search and identify unique foods (legumes, roots, nuts and berries) in terrains that had unusual topography. Males had to cooperate with other males in hunting and scouting parties if a successful harvest was going to be made possible. Males within each small group or clan quickly learned that if they did not work together collaboratively in their efforts to secure food, they would all suffer (and more than likely ultimately perish) unless they worked as a team. This would mean not only tracking and hunting big game with the maximum amount of protein, but also breaking up into groups to help drive game herds into an area that maximized chances of harvesting the food.

Women also evolved with the capacity to work collaboratively but in different ways. Rather than having one physically stronger person (i.e., the Alpha male) delegating responsibilities and dominating how group members worked with each other, women tended to work more interdependently and shared responsibilities in securing and preparing foods. Females also evolved with memory capacity to identify and locate these areas for future use (Eals and Silverman 1994). This sexual division of responsibility and labor produced distinct gender differences regarding skills and aptitudes vital for survival within a variety of different environments. Men and women who were most successful in producing foods and resources provided the greatest reproductive success in future generations, thereby achieving significant sex-differentiated traits in future populations.

THE ETIOLOGY OF GENDER DIFFERENCES: ARE WOMEN MORE COOPERATIVE THAN MEN?

The sociocultural argument in describing human behaviors argues that any differences noted in behaviors among men and women are due to factors relative to the culture from which one is exposed to as well as characteristics that are learned and rewarded as individuals mature and development. Thus, gender differences (at least according to the sociocultural view) are essentially learned through a complex interaction between one's own culture and experiences that are shared during development. This view is in sharp contrast with the theory that gender differences exist as a function of evolutionary and biological changes that they have according over the span of several million years.

The evolutionary argument suggesting distinct gender differences in skills relative to survival (i.e., hunting and gathering as opposed to foraging and recalling vital food sources) is in direct contrast to other more popular views addressing the etiology of gender-related differences. For example, in Maccoby and Jacklin's (1987) classic study addressing gender-related differences, early childhood same-sex interaction accounts for differences in behaviors among men and women later in adult development. Specifically, Maccoby (1990) has argued that girl's same-sex interaction accounts for a greater display of prosocial and cooperative behaviors among adult females. Stated simply, the primary distinction between the sociocultural and evolutionary perspectives in gender-related cooperative behaviors is that one is learned (sociocultural) and the other is inherited through distinct adaptive practices early on in our evolutionary history.

Eagly and Wood (1999) argue that an important reason why we see gender-distinct styles of behavior has more to do with culture and the value of social roles placed on women (i.e., domestic work, cooperative group behaviors, monogamous relationships and child-rearing activities) within these cultures. It is not surprising, then, to see more progressive cultures show a more egalitarian and equal displacement of men and women sharing a broader range of both professional and domestic responsibilities in the work force.

In Maccoby and Jacklin's (1987) classic research exploring what factors may have contributed to gender-related differences in a variety of different types of behaviors (i.e., cooperation), the argument that differences in economic structures and hierarchies provide the basis for differences in a variety of behaviors among men and women. Furthermore, Maccoby and Jacklin argue that if society were more equal relative to economic opportunities and domestic responsibilities, then there would be a significant reduction in gender-related behaviors (i.e., cooperation, empathy, and aggression) (Eagly and Wood 1999).

The evolutionary argument that explains gender-related differences in behaviors is not based on societal structure, sociocultural theories or economic principles, but rather how men and women adapted to a variety of environmental problems throughout our early evolutionary history. Sex-differentiated adaptations provide the fundamental basis of behaviors among men and women within a myriad of complex environments. Interestingly, as sex-differentiated skills evolved over time as a means of improving resilience and survival skills (i.e., women tend to outperform males relative to spatial recall of a variety of different objects and structures), adaptive behaviors (i.e., aggression and cooperation) also evolved with unique sex-related differences:

For example, when males engage in a variety of tasks with other males, their ability to remain highly cooperative and interactive significantly increases, but as the groups become more heterogeneous (i.e., women are present), their levels of cooperation significantly decrease as a means of establishing male hierarchy (Bowles 2006). In this way, the gender of the makeup of the group has played a significant role in how individuals interact with each other. A homogeneously male-dominated group would show the highest levels of cooperation (Bowles 2006), however, as the group becomes more sexually heterogeneous, males show significantly lower levels of cooperation and higher levels of competition.

According to current evolutionary theory, the gender context of specific groups determined how problem-solving strategies emerged. Males tended to work cooperatively when working with other males in a hierarchical fashion (Balliet and colleagues 2011; Rapoport and Chammah 1965) until the gender balance changed involving more women. Conversely, evolutionary theory argues that women tended to behave more socially and communally within a variety of situations involving social dilemmas, and this interactive behavior even extended to strangers or individuals with no prior interaction (Wood and Eagly 2010).

Finally, because of environmental demands and physiological differences between men and women, women have traditionally displayed a stronger capacity for interpersonal engagement and interaction and (at least perceived to be) less selfish, more cooperative and emotionally descriptive in their interaction with others in social environments (Eagly 2009).

EVOLUTIONARY PERSPECTIVES ON SEX-DIFFERENTIATED SEXUAL STRATEGIES

Individuals may have differences in opinion regarding sexual attraction and potential mates, but current evolutionary theory argues that men and women have distinct preferences that aid in the development of reproductive fitness. Darwin (1871) and David Buss (1989) note that humans are both ingenious and highly adaptive creatures when faced with critical decisions facing their own survival. Additionally, both Darwin and later Buss have also noted that given the highly polygynous nature of humans, their ability to reproduce was essential in their efforts to survive particularly harsh and unforgiving environments. The infant mortality rate among humans during the evolutionary period dating over 150,000 years ago (although some evidence suggests that Neanderthals existed as recently as 35,000 years ago in Croatia) was extremely high, where typically only one in ten (1:10) infants were expected to survive past their 10th birthday, and the average lifespan for the Neanderthal was ranged between 20 and 27 years.

In a classic study that explores specific gender-related differences in relative to jealousy, Buss, Larsen, Westen, and Semmelroth (1992) found that males report the highest levels of distress in a hypothetical romantic relationship where they were asked to think of their partners as being either emotionally or sexually unfaithful. Over 60% of males reported greater distress over their partner's reported sexual infidelity, whereas only 17% females reported greater distress over their partner's reported sexual infidelity. Conversely, 83% females reported greater distress when their partner was perceived to be emotionally attached to another female. The primary distinction between men and women was the perception of between "sex" and "love."

Consistent with traditional evolutionary theory addressing male and female sexual strategies, males show greater distress over perceived sexual infidelity and females greater distress over emotional (i.e., "love") infidelity. A female that was impregnated by another male during our early evolutionary history and whose offspring was raised without any biological or genetic kinship would have been disastrous for males (i.e., "cuckolded") and ended his reproductive lineage. Similarly, females risked losing their partners if the male became emotionally invested with other women during our earlier ancestral period, and this abandonment could lead to disastrous consequences as well relative to child protection and providing resources to her offspring.

Traditional evolutionary theories have made strong arguments for distinct mating strategies and specific sexual behaviors among males and females (Reiber and Garcia 2010), where males have historically engaged in higher frequency of sexual behaviors with a broader variety of mates (fertile females) ultimately leading to a higher reproductive success rate (Daly and Wilson 1983). Conversely, women's sexual behaviors (at least from the evolutionary perspective) have traditionally remained more selective and persnickety when identifying potential mates (Buss 1989), and for good reason. According to the evolutionary perspective, the advantage for women who select an ideal mate has to do with offspring and resources. A physically stronger and more resilient male can not only provide more resources for her offspring, but can also provide greater amounts of protection in dangerous environments. As Buss (1989) notes, these skills are particularly important for women as they become more incapacitated during childbearing times. Additionally, males who were more concerned about their own reproductive success and concerned about phylogenetic progeny relied more on aggressive-related behaviors to ensure that they were raising their own offspring (i.e., preventing "cuckoldry").

There is numerous empirical evidence that strongly supports the theory that both men and women displayed highly differentiated sexual strategies during evolutionary history, and that these differences and preferences in mate selection exist even today (Reiber and Garcia 2010; Trivers 1972; Clutton-Brock and Parker 1992). In a recent study that explored how college students today sexually interact with each other, Reiber and Garcia surveyed over 500 undergraduate students and asked them explicit questions pertaining to "hooking up." Hooking up was operationally defined by the authors as: "Engaging in a no-strings-attached (NSA) sexual behavior with any uncommitted partner (p. 390)." Specific types of behaviors were divided into five different categories (i.e., sexual touching above the waist; sexual touching below the waist; oral sex performer; oral sex receiver, and intercourse) and scores were measured via a Likert scale (+5 very comfortable to –5 very uncomfortable). Overall, male respondents indicated feeling more comfortable with oral sex (receiver) and intercourse ($M = 2.91$, $SD = 3.19$) when compared with female responses ($M = -1.99$, $SD = 3.68$). Additionally, the authors concluded that the female respondents in the current study attributed to men higher comfort levels. What this means in simple language is that women overall felt more uncomfortable with the concept of a "NSA" encounter and felt that men were more likely to be comfortable with a variety of uncommitted sexual partners.

The results of the current study strongly support the evolutionary premise that men and women have very different sexual and mating strategies which have been attributed early on to our evolutionary past. According to the authors of the current study, males were significantly more comfortable with a variety of sexually related behaviors (i.e., "NSA") than females, and the evolutionary perspective strongly supports this as a highly effective strategy in improving one's own reproductive success and phylogeny. Interestingly, the study also identified males as actually having an accurate awareness of women's discomfort with sexually related behaviors.

Numerous empirical studies have supported the theory that significant gender differences do exist regarding sexual activity and mating strategies and that while cultures may influence differences in sexual strategies and mating patterns, there is near universal agreement in the characteristics and motivational factors that address sexual encounters and relationships among men and women. There is little doubt among scholars today that men and women do have significant sexual strategies when identifying and selecting potential mates (Gangestad, Simpson, Cousins, Garver-Apgar, and Christianson 2004); however, the real controversy arises over the etiological factors in these gender differences.

Evolutionary theorists argue that males have historically preferred short-term mating practices as a means of increasing reproductive fitness as well as exemplifying social dominance among diverse groups (Sundie 2011). Furthermore, empirical evidence shows that while women historically have preferred monogamous long-term relationships with the same mates over short-term polygamous relationships, females have also shown preferences toward dominant males as a means of protection of offspring and vital resources. In summary, then, while cultures may vary in terms of the traditional rituals involved sexual and mating strategies, there is striking universal similarity in *why* men and women have become sexually attracted to one another.

MATE PREFERENCE, "PERSNICKETIES" AND EVOLUTIONARY THEORY

Differences in preferences of physical or phenotypical structure and body types have been shown to have an important evolutionary background (Buss 2012). What individuals report to find sexually attractive and appealing is certainly more than a "skin deep" phenomenon. Even heterosexual males who happened to be born blind showed preferences to mannequin dolls with a .7 WHR as compared to mannequin dolls with different WHR's (Karremans, Frankenhuis, and Arons 2010). Human preferences to finding sexually attractive and compatible mates and partners have evolved over significant periods of time as a means of improving sexual and reproductive fitness. Cues of sexually attractive potential mates, such as shoulder-to-hip ratio and waist-to-hip ratios, as well as skin texture, visible musculature and body fat characteristics all proved to be accurate predictors of ideal and robust physical health that determined overall reproductive fitness. It seems that Darwin was correct in postulating that the purpose of life itself is through sexual reproduction.

Researchers addressing the complex issues involving sexual attraction and mate selection have determined several key factors that are important and relative to evolutionary theory, such as parental investment theory (Trivers 1972), social status and financial resources (Buss and colleagues 2001). Females biologically have more reason to be selective regarding potential mates. For examples, males produce sperm (approximately up to 12 million per hour) up until late adulthood (many males well into their seventies and eighties have sired children), whereas females produce a limited number of ova during their lifetime (approximately 400). Additionally, women are tasked with breastfeeding offspring and general child care such as feeding and protection. According to Trivers' parental investment theory, because women are typically tasked with more responsibilities during pregnancy and child-rearing, they needed to be highly selective in choosing those mates who would be best equipped to provide for them through financial resources, status and specific personality traits.

Males providing a broad selection of resources to prospective females have historically held an advantage in attracting fertile females for reproduction. This preference is not limited to a particular group of women but has been noted by Buss and colleagues (2015) to exist in over 37 different cultures. Social status has been considered to be important to females because high social status represents one's ability to *access* important resources (i.e., food, housing, etc.) is significantly improved. Personality traits (i.e., humor and kindness) as well as behavioral tendencies (helpfulness and ambition) have evolved for women as important evolutionary traits to mate selection. Mates who show a greater willingness to work harder and succeed are significantly more likely to be economically successful and thus provide greater resources for their families.

During our evolutionary history those women who were less discriminate regarding potential mates had a significantly lower reproductive success rate. Conversely, those women who showed higher levels of "mate persnicketies" (a tendency to be very selective in potential mates based on personality traits and behavioral tendencies) showed significantly higher rates of reproductive success. These evolutionary traits still remain with women today in helping them select mates best suited toward child-rearing and adaptation. The processes by which both males and females have selected their mates have been influenced over long periods of time through a variety of evolutionary-related themes, including reproductive success, adaptation and overall survival skills.

Chapter 8

The Evolution of Language Relative to Cooperative Human Behaviors

Few mammals are born as dependent and helpless as those of humans. Without immediate care, protection and nurturing, infants would quickly succumb to the demands of the outside world. One very important skill relative to human survival is that of language capacity. The development of language skills is a very important topic within the discipline of evolutionary psychology because language presumably evolved directly as a function of helping humans not only to adapt to the demands of the environment, but perhaps even more importantly is the fact that language development helped humans coexist and function cooperatively in small groups.

What exactly is language and how did the development of language contribute to the development of human civilization and cooperation? One of the most complex questions to ask relative to human evolution is that of communication, or more specifically the development of languages within the context of human development. A related area of study that explores the scientific study and origin of languages and how culture influenced the development of language is evolutionary linguistics. We will explore the reciprocal relationship between language acquisition, interactive cooperative behaviors and a process of learning described by Vygotsky (1978) as Zone of Proximal Development (ZPD). Language development facilitated cooperation among early human ancestors and cooperative exchanges facilitated the development of languages. As a final note, in this chapter we will explore how languages actually evolved and the role of language in establishing human relationships and cooperation.

THE CASE OF GENIE: A FERAL CHILD WHO NEVER LEARNED LANGUAGE

Imagine a 13-year-old child who had never been exposed to any sounds in any language and was literally locked away in her bedroom her entire life. Sadly this true story began to unfold in Arcadia, California where "'Genie'" (not her real name) was discovered by social workers and rescued after years of neglect and abuse (Rymer 1993). Genie was discovered by social workers in November 1970 when her mother inadvertently entered a social services office after an argument with the father. An alert social worker sensed that Genie was a victim of neglect and contacted the local authorities to conduct an investigation. On November 4, 1970 Genie became a ward of the court and was admitted to Children's Hospital where intense psycholinguistic and neurological therapy began. Her case presented a unique opportunity to test the theory if language was possible of being developed after years of neglect. Previous to the case of Genie, language experts argued that even with minimal exposure to sounds during development some form of language could be taught to children who had been neglected or abandoned (Curtiss 1977). After several years of intense training and psycholinguistic intervention with some of the finest psychologists and medical doctors in the United States, Genie unfortunately was never capable of learning basic language skills.

Increasing empirical evidence supports the theory that the evolutionary capacities of language and communication have developed within the context of group interaction that served several needs during human development (Cartwright 2000; Blurton-Jones 1972). In an earlier study (1976) conducted at the University of Minnesota, researcher David Johnson and colleagues discovered that *how* children are taught language skills can have an important impact in not only what they are learning but also influence perspective taking of others that resulted in more altruistic behavior among children (Johnson, Johnson, Johnson, and Anderson 1976). When children are taught language skills within a cooperative manner (i.e., group work where ideas and suggestions are shared in completing each assignment), students were more accurate in identifying how their peers were feeling (empathy) and were also able to complete a variety of language assignments with fewer errors than the control (individual students) group. It appears that when we are able to work cooperatively and collectively within small groups we are better able to share our own ideas with others as well as identify how individuals respond to those ideas in completing assignments.

According to teachers who were associated with the original study, students in the cooperative group were also better equipped to follow instructions and experienced less difficulty in completing assignments. The results of Johnson and colleagues (1976) as well as the case history study of Genie suggest that learning languages is clearly a time-sensitive phenomenon (i.e., critical period) that can also be enhanced when taught in specific modalities, such as in small groups or cooperative design (Johnson and Johnson 1975).

THE EVOLUTION OF LANGUAGE: WHY DID LANGUAGES EVOLVE?

Different theories exist in explaining the reasons why languages have evolved. Two competing theories offer very different views regarding how languages have evolved. Some of the more recent theories in evolutionary psychology have argued that the development of languages among humans as a species evolved specifically as a means to facilitate adaptation and survival (Pinker 1994; Pinker and Bloom 1990). Humans required some form of communication as a means to adapt to the increasing demands of the environment (i.e., fending off predators) as well as a tool to facilitate social interaction and cooperation among small groups. Additionally, languages share several cross-cultural similarities such as verb structure, syntax and grammatical rules in general that suggest the development of languages during human evolution was by design and not an incidental biological phenomenon.

More traditional arguments within the discipline of psychology have argued that the development of the brain (and other physiological developments) exists through the processes of natural selection. Noam Chomsky's classic view of the Language Acquisition Device (LAD) and the development of universal grammar both support this view. Given the tremendous potential that the human brain presents with all of the processes involving synaptic transmission, neurons, etc., language emerged with other functions (i.e., intelligence) and related cognitive skills during human development.

Few human behaviors are more important than the role of language, yet ironically language is really something that many of us take for granted. Language can take on many forms—either verbally, written or signed, and each has played an important role in shaping the way humans interact and coexist with each other. Humans developed the capacity for language approximately 100,000 years ago, when grunts and hand gestures evolved into primitive sounds and short sentences that gradually took on meaning to express thoughts and ideas. All languages began with very short, basic utterances that gradually came to represent meaning in more elaborate sentences. The term "'phonology'" refers to collection of different types of sounds that combine to form meaning. Vowels and consonants merge to create a unique communication system that we refer to as language. An infant who sees a bright red apple and hears the combination of sounds to produce the word itself will typically imitate those sounds emitted by the caregiver. All senses (i.e., vision, touch, smell, hearing, taste) become incorporated in the effort to produce sounds that gradually merge into a formal language (Kuhl and Damasio 2011). For example, a child sees the object that he or she is trying to verbalize, hears the sounds necessary involving phonology to pronounce the word, and perhaps even “feels” how words are pronounced by touching the lips of the caregiver while he or she is making deliberate sounds of each object.

The development of language then is clearly a reciprocal process that involves mutual attention and interaction between caregivers and infants in mastering basic rules necessary to begin learning the skills needed in language development. Phonemes refer to the basic sounds contained in any language (i.e., vowels and consonants). An infant sees a red apple that his mother is holding and hears the word “apple.” While repeating the exact word may be too complex at his age of 12 months, he is capable of imitating vowels (“AP”) and consonants (“PULL”) and soon will master the ability to combine these sounds into “Apple.” Morphemes refer to the smallest unit of meaning within languages (i.e., prefixes and suffixes), and syntax refers to differing sets of rules that govern how languages are used among any group of people. Early on in development, infants may typically gesture (i.e., point) to objects in illustrating their intention to communicate, and these gestures develop into short grunts or squeals that may be referred to as babbling. These sounds are all used in an effort to not only communicate to others in a primitive fashion, but they become the actual building blocks of language in future development (Lock and Zukow-Goldring 2010).

Recent research also suggests that how successful an infant is in mastering language is actually more of a combination of evolutionary and genetic factors (Pinker 1994) as well as environmental stimulation and exposure to the sounds and frequencies in which words are emitted (Woodward, Markman, and Fitzsimmons 1994). For example, even with limited exposure to words, children seem to be able to comprehend the basic nature of sounds and syntax of words, a phenomenon that is referred to as “fast mapping” (Woodward, Markman, and Fitzsimmons). Activities that involve early literacy programs as well as reading to children and exposure to vocabulary at early ages all seem to have very positive impact on the rate and quality of learning languages among children

How humans may have communicated with each other early in our evolutionary history remains a highly complicated subject for several reasons. Perhaps one of the most problematic issues is simply what we do not know relative to early human evolution and language development. We do not know precisely when languages evolved in primary use for communication among humans, and we do not know when early humans relied less on physical gestures and basic or primitive grunts as communication tools and when a more sophisticated form of language came into use. Furthermore, we scientists are unclear when humans began using languages in a way that facilitated their own existence in the modern world. Indeed, many early scholars considered the problem to be essentially an "unanswerable" one and simply chose not to explore the discussion further (Christiansen and Kirby 2003).

Despite these formidable problems, we can and should explore the process of how languages evolved and its impact on the development of communities that required cooperative relationships as a means of survival. Simply because a particular topic (i.e., understanding how and when human languages evolved) is complicated and difficult should not warrant or justify scholars to abandon their study, but rather should serve as a catalyst for rigorous academic study and research. Most scholars and evolutionary linguistic experts agree that not only do humans possess an innate capacity for languages (something that historically has made our species distinct from other species), but humans *need* to be exposed to environments that stimulate speech development for healthy cognitive and neurological development. Recent evidence in evolutionary psychology has determined that specific components of the cerebral cortex evolved in relation to the human capacity to speak languages (Friederici, Mueller, and Oberecker 2011). For example, the area known as Broca's area (located in left frontal lobe) and Wernicke's area (located in left hemisphere of the brain) both contribute to speaking and understanding languages. Significant organic damage to either area increases the likelihood of an inability to speak language (a condition known as aphasia).

One of the first sub-human species to walk erect and display distinct physical characteristics that separated it from lower primates was *Australopithecus* or more commonly known as "'Lucy.'" Most scholars agree that Australopithecus existed in eastern portions of Africa some 4 million years ago and played an important role in the development of human cognition and language development. Although scholars agree primarily that Australopithecus lacked a formal language today as we know it, most scholars also agree that Australopithecus were the first species to show evidence of genes that contributed to neural development within the frontal lobe, an area known for language comprehension and development (i.e., Wernicke's area and Broca's area).

Additionally, evidence shows that while the Australopithecus lacked a formal language, there is evidence that the Australopithecus were capable of bipedalism (i.e., walked upright), and worked cooperatively in small groups as upright posture provided the capacity to explore above tall grasses or trees to identify possible predators and identify viable food sources (Thorpe, Holder, and Crompton 2007).

This physiological development, together with the evolution of cooperative behaviors within small groups, was remarkable given the significantly smaller cerebral cortex of the Australopithecus (the brain was only 35% of what modern humans have today). In other words, while Australopithecus may not have had any formal language other than simple grunts and gestures, they did provide for us the genetic components as well as the fundamental skills to work in small groups that later would enable a more precise language to formally develop. Additionally, the Australopithecus species would later evolve into a more modern species similar to Homo sapiens or modern Cro-Magnon approximately two million years ago.

CONTINUOUS VERSUS DISCONTINUOUS THEORIES & THE DEVELOPMENT OF THE HYOID BONE

Two primary and very distinct schools of thought emerge when identifying how languages evolved in human history. They are described as the "Continuous versus Discontinuous" theories and are essentially very different from one another. In the former (continuous) theory, languages are described as highly complex tools of communication, where languages must have continuously maintained their development throughout the history of human evolution. Languages are described through a gradual (i.e., continuous) series of developments over long periods of time, probably from a pre-lingual era that first involved grunts, gestures and very basic and primitive sounds. The early beginning forms of languages were necessarily very primitive and rudimentary, and over time as human evolved and developed cognitive and intellectual skills, so did languages reflect the complexity of human evolution. Scholars who argue that the development of language is primarily due to a system of rewards and punishments (similar to the Behavioral approach) would support the continuous theory of language development. When infants begin to practice uttering basic sounds (such as vowels or consonants), the caregiver engages in direct interaction with the child, providing positive reinforcement so the child continues his or her effort in language mastery. The relationship is reciprocal, as the child gleefully interacts with the caregiver and the caregiver smiles and rewards the child via praise for making efforts to pronounce words: "Ma–Ma."

Conversely, the discontinuous theory of language argues that language development did not occur gradually over time and that because of the inherently unique complex requirements for the development of language, it must have occurred relatively rapidly as a result of the emergence of some gene or structural development in the larynx (i.e., the hyoid bone). Theorists such as Noam Chomsky support the Discontinuous theory of language development and argue that the evolution of some specific gene or trait must account for the development of modern languages more than 100,000 years ago.

Simply stated, Chomsky (1957) argues that the introduction of something as advanced and complex as languages over 100,000 years ago must have been genetically unique among humans. Languages are much too complex to have evolved gradually over time. Furthermore, Chomsky argues that all humans possess an innate language development device (what he refers to as the language acquisition device or "LAD") where humans need to be exposed to enriched environments that will trigger or stimulate the process of language development.

This inborn language capacity (at least according to Chomsky) is unique among our species and has had profound influences in the process of the development of civilization and cooperation among groups within various societies and cultures. More recent study in addressing the capacity of the lower primates to learn at least a rudimentary form of language (including sign languages) has become controversial in light of the research offered by Penny Patterson (http://en.wikipedia.org/wiki/The_Gorilla_Foundation).

The Discontinuous theory of language development sees languages first emerging probably several hundred thousands of years ago, most likely with the emergence of Neanderthal species and a vital component to language in the larynx that is referred to as the hyoid bone. Additionally, the Discontinuous theory of language development argues that the development of language itself is necessarily genetically encoded from human evolution (most likely at least from Australopithecus) and therefore is a unique function among humans.

LANGUAGE, NEANDERTHALS AND THE EVOLUTION OF THE HYOID BONE

The Neanderthals were the first species of humans to have evolved within their throats as a key component that is necessary for making distinct sounds (phonemes) that are typically characteristic of modern languages. This device that evolved in the Neanderthal's throat is referred to as the hyoid bone. Most modern scholars and psycholinguistic experts consider the evolution and development of the hyoid bone to mark the beginning of the first use of formal languages among humans (Arsenburg, Schepartz, Tillier, Vandermeersch, and Rak 1990).

Conversely, the discontinuous theory argues that languages developed rapidly as human societies and civilizations were created. The discontinuous argument states that languages evolved as precise and complex forms of communication that simply could not have gradually or slowly evolved. Many scholars today believe that the Neanderthals were the first to possess the hyoid bone and from this key structural development the larynx gradually developed and allowed for the development of formal modern languages. The presence of the hyoid bone provided opportunities for the development of more sophisticated languages and thus supports the argument of discontinuous language development.

The hyoid bone was first discovered in the Kebara Caves of Israel in 1982 under the supervision and direction of Prof. Ofer Bar-Yosef. The caves were composed of limestone deposits, and researchers agree that they were inhabited by the Neanderthals approximately 60,000–48,000 BP. Additionally, as the hyoid bone gradually evolved and facilitated the development of language among the Neanderthals, the ability for smaller groups or clans of individuals provided for the development of more sophisticated cooperative behaviors. The evolution of the hyoid bone also provides further support for the discontinuous theory of language development as the larynx quickly provided opportunities for Neanderthals to formalize developed languages through the use of phonemes and morphemes.

BEHAVIORAL THEORIES OF LANGUAGE DEVELOPMENT

The process of learning and adaptation among human development has remained the central theme of this text. Many current (and earlier) psychologists have argued two primary themes when addressing how languages have evolved. The two primary arguments address the Behavioral approach and the Nativist theories. The Behavioral approach and the Nativist approach are not unlike the classic "'nature–nurture'" controversy in psychology. It would be helpful to bear in mind especially as you review this chapter that human behavior (especially discussing topics such as language development, intelligence and cognition) can never really be separated from environmental factors. Human behavior is a complex interaction between our biology and genetic constitution and the kinds of physical experiences that we have had in our environment. So learning a language then is really about exposure to different types of environmental stimuli and our own genetic makeup.

If you may recall from previous chapters, the Behaviorists argued that all aspects of human behaviors are essentially functions of the interactions between people and their environments. John Watson (and later B. F. Skinner) argued that the single most efficient way to understand an individual is to not explore the person, but to explore the environment and child-rearing habits that they have been exposed to. According to these well-known Behaviorists, the best way to process information and learn language is through reinforcement or conditioning process. Thus, according to the Behavioral approach to language learning, reinforcement and interactive processes that stimulate learning in a positive way are the most effective methods to teach children to learn a variety of different skills in language development. If parents praise and reward their young children for attempting to imitate a particularly complex word, then the child is more motivated to continue trying through trial and error until pronunciation is proficient. Thus, language development is essentially about building a relationship between child and parent and establishing a system of rewards and punishments as a means of developing an advanced form of language. This process is not unlike learning other types of cognitive activities in human development.

Learning a language, according to the Behaviorists, is no different than learning other things in our life. If a parent wishes to teach a child a language, the first process involves interaction and a system of rewards when the infant first begins to babble vowels and consonants. At this point (usually about 3–6 months), infants have the capacity to utter any sound in any language—this period is often referred to as the "critical period" of language development.

In his classic 1957 text: *Verbal Behavior*, B. F. Skinner offers parents direct advice and information regarding the most efficient way in teaching infants the beginning mechanics of language development. When infants first begin to utter their sounds, it is critical for parents to reward (i.e., provide attention in the way of smiling, direct eye contact, and physical interaction) which will stimulate the infant to gradually make more complex sounds that serve as the building blocks of languages. The infant continues his or her efforts at making sounds because of the preferences of the increased interaction that the parent provides. Skinner provided a summary of the specific variables that stimulated and promoted language development.

THE NATIVISTIC APPROACH TO LANGUAGE DEVELOPMENT

Unlike the Behavioral approach to language development, the Nativistic approach argues that humans are distinct and unique creatures in cognitive skills that enable them to master any language. An early proponent of the Nativistic approach to language was the famous psycholinguistic expert Noam Chomsky (1957). Earlier in this chapter we discussed how Chomsky argued that humans have an innate capacity for language and come into this world biologically "'prewired'" (i.e., readily capable at birth) to begin speaking languages. The key to successful language development is an interaction between environment encouragement and stimulation from caregivers as well as our ability to respond to natural language skills that unfold over time. The process of the Nativistic approach to language development is considered to be discontinuous because of the recent genetic evolution of FOXP2.

The FOXP2 gene is actually a protein found as recently as 300,000 years ago in some bone deposits of the Neanderthals from the Denisova Caves located in Siberia, Russia.

However, while all humans are in fact capable of speaking a broad range of different languages utilizing different sounds (i.e., phonemes), humans also have a limited window of time to master these unique sounds in all languages. Non-exposure to unique sounds in different languages makes it almost impossible for humans to imitate and duplicate these sounds later in adult development. The fact that humans have difficulty in adapting to unique complex phonetics and sounds in various languages later in adult life suggests that mastery of polysyllabic skills requires more than simple reinforcement principles. The Nativistic approach to language development among humans argues that humans were not always capable of modern language, and that some radical genetic development must have occurred that enabled humans to develop and process language in a meaningful way.

The Nativistic approach is discontinuous in the sense that learning occurs only with the onset of specific physiological and genetic development (i.e., the hyoid bone in the larynx) as well as the development of the gene FOXP2. Thus, the Nativistic approach argues that innate skills and genetic capacities differentiate humans from the lower primates in our ability to speak a variety of languages. For example, while many animals and lower primates may have advanced communication systems and even sign language capacity, human languages have distinct characteristics including: Phonetic pronunciation, productivity (humans can typically create indefinite sounds and utterances), semantics, and displacement (communicating ideas in the abstract whereas lower primates lack this skill). Many linguistic scholars differ in their opinions when humans first began to speak languages. Interestingly, some linguistic scholars also argue that humans are not the only species capable of mastering some form of a language, albeit a rudimentary and very primitive form of a language (Fitch, Huber, and Bugnyer 2010).

Previous discussions addressing how evolutionary theory has influenced human existence have identified key components relative to how humans have survived in particularly harsh conditions. One continuous theme has remained, however, that provides strong support in how humans have been able to show resiliency and adaptation in light of these challenges. That theme remains to be the context in how humans interact, communicate and cooperate with each other. Language acquisition and development would have been impossible to develop within our early evolutionary history without the existence of some form of positive interaction and cooperation among humans within small groups.

This chapter has explored the development and basic roots of language and whether or not humans are unique in their capacity to use a formal language as a means of facilitating our own survival. Many researchers and experts in human evolution would agree that while lower primates may have advanced communication skills, only humans possess an advanced language with syntax. The question remains, however, in early human evolution how languages even began to develop—at some point, small groups or clans of human ancestors must have agreed on some systematic form of rules (i.e., grammar) and pronunciations (phonemes) so they could at least understand each other within a rudimentary fashion. Individuals residing within those smaller groups or clans needed to cooperate and engage in some formal system of sounds, pronunciations and words that would make their own existence clearly more capable of survival in an environment with many challenges. Furthermore, early humans (most likely as recently as 300,000 years ago with the Neanderthals) needed to create a system where their thoughts and ideas (i.e., how to harvest foods, protection from predators, etc.) would be capable of being transformed into some primitive language.

Famous Russian Psychologist Lev Vygotsky (1896–1934) introduced an important concept relative to understanding the cognitive world of children and developmental psychology. Similar to Chomsky's Nativistic theory of Language Acquisition Device, Vygotsky argues that children have an inborn mechanism that stimulates the capacity and competency to master a language. Vygotsky argued that as children begin to mature, they develop the capacity to transform images from their environment into symbols, and that languages help to convey meaning from these symbols into a communication system. One effective way in which children master languages (as well as other cognitive and physical tasks) is through the concept of ZPD. Simply stated, the ZPD is the difference between what a child is capable of doing and what they potentially can do with the assistance of adults or a mentor. Without the cooperation of groups of adults in encouraging and teaching children how to formulate sounds into vowels and consonants, the development of a formal language would have been impossible. A two- or three-year-old child may attempt to formulate specific sounds by experiences difficulty in doing so. The adult model facilitates the process of making distinct sounds by showing the child how to formulate his lips ("Peach" with lips expressing the distinct "P" sound), or the adult places the child's fingers over his or her mouth as they pronounce "Peach" so the child may imitate (however crudely) the sounds necessary to duplicate the word "Peach." This distinct style of individualized of adult teaching and interaction with younger children has been described as "scaffolding."

Although the term scaffolding was never actually used by Vygotsky, the term was actually introduced by Wood, Bruner, and Ross (1976) as a metaphor used to describe the importance of interaction between adult models that would encourage children to formulate complex sounds during the process of language development. The process of scaffolding requires cooperation among not only the child with the adult model, but also with other children during the process of development. Vygotsky argued that language development was also a social process where children learn to process words and sounds as their ability to create symbols also develops (i.e., assimilation). Similar to Piaget's theory of assimilation, the child incorporates meaning into his or her vocabulary and the process becomes more elaborate and precise as the child's experiences in their environment necessitates the expression of more complex sounds and syllables that are incorporated into their language.

The ZPD has been described as a fundamental concept relative to the development of language because of the distinct and necessary cooperative relationship between the child and the adult caregiver. Vygotsky argued that the ability for a child to master and learn a language depends on stimulation and interaction within the environment in which the child is exposed to. Additionally, Vygotsky would agree with both Chomsky and Piaget that humans are distinct organisms when mastering languages because humans have an innate (but limited) timeline and capacity for exposure to key sounds that require practice and rehearsal. Furthermore, as the capacity of the child's ability to speak languages becomes more elaborate and advanced so does his or her cognitive skill—advanced speech stimulates advanced thinking capacities for both children and adults. Adults who simply ask more "open-ended" questions such as: "Why do you like playing with animals?" will stimulate children to think in more in-depth and abstract ways which has also been shown to trigger the capacity for extended language development (Bates 1976). The ZPD, therefore, can best be described as the zone or area between what children can do and what they are *capable* of doing. The capacity of aptitude and performance among children is significantly increased within a supportive and educational environment among group members.

TEACHING LANGUAGES VIA COOPERATIVE LEARNING THEORY

In the final section of the chapter we will review the reciprocal relationship between the evolution of languages as a communication process and cooperative behaviors. Languages (described here as some form of symbolic communication among groups of individuals) have evolved through the cooperative behaviors throughout our evolutionary history and similarly the development of cooperative behaviors has helped to formalize and develop modern languages. Initially we may speak in very basic and primitive terms, but with the collective efforts of others within the group we slowly begin to understand the concepts and terms that are used in the language we are trying to understand. Similarly, our relationships with those who are helping us master communication also changes, typically in a very positive way.

We will begin our discussion by exploring the psychological impact of groups and languages, and how individuals within the community can often help others master and learn the language they wish to speak. Much research has been addressed that explores the role of language in resolving conflicts and problems that are common in our society (Smith 2010). Languages have also been shown to be instrumental in resolving unique types of human problems involving social dilemmas, where one person's gain may come at another person's (or groups') loss.

Learning any language is not an individual or solitary event. Even adults who learn a new language often comment how much faster and easier it is to process new languages when they are exposed to the culture and people who are speaking a foreign language. Unique problems develop within any culture or society where language and communication become misunderstood and isolated. Certain types of problems emerge that require all members of the group to work together to find various solutions. For example, employment, education and affordable housing are very common problems facing variety of societies and these problems become even more complicated when the issue of language is not entirely understood by all members of the community.

Smith (2010) refers to these kinds of problems as "collective action problems" or CAP's which are by design very similar to the prisoner's dilemma discussed earlier in chapter 5. Smith argues that languages can play an instrumental and effective role in the resolving conflicts that typically involve maximizing individual gains over society or the group's best interests. Languages that are more advanced and available to larger populations can be used to identify specific norms and rules within various cultures and thereby reduce the probabilities of conflict among individuals. Human communication is unique in the sense that it allows for a broad range of problem-solving capacities and what he refers to as "collective action" in addressing a variety of problems that have faced individuals throughout our evolutionary history. Collective action has been described by Smith as "any type of situation where individuals must work cooperatively as a means of producing something valuable or good relative to human welfare". This may mean combining skills as a means of harvesting food or warding off some kind of threat or danger.

Concepts relative to cooperative learning and language development argue that an individual's capacity to learn any language is not through an isolated or separate process, but rather through social interaction. Cooperative learning that involves language development occurs through groups of individuals that requires some exchange of information. This information is shared by members of the group and provides a mechanism to identify the elements that are processed and learned, and stimulates individuals to continue learning and sharing information.

Cooperative learning has also been shown to develop other positive psychological qualities within the group, such as positive interdependence and social skills (Slavin 1990). Many primary level educational programs today have structured their programs on basic cooperative learning principles. Researchers (Skon, Johnson, and Johnson 1981; Johnson, Johnson, Johnson, and Anderson 1976) have identified cooperative learning mechanisms as a central feature that facilitates language development when combined with key components, such as direct eye-to-eye contact with the learner, shared reading skills, exposure to new and unique sounds central to the language being learned, and slow, articulate word pronunciation. Cooperative learning principles have also been shown to help children learn social skills when interacting with individuals from different cultural and ethnic backgrounds as well as understand perspectives from other people.

Some examples involving language development through cooperative learning mechanisms include the "jigsaw method" (Slavin 1990), "Think-Pair-Share" method and the "Structural approach" (Kagan 1992). Each approach that is used to teach language through cooperative mechanisms involves social interaction (i.e., children exchanging information with each other) as well as the development of positive interdependency. Interdependency is characterized as positive meaning that all members of the group advance but only when individuals of the entire group collectively work together (i.e., no social loafers).

The jigsaw approach is used in many ways by educators of various levels that require all group members to contribute in some way. Work assignments are divided into sections and members contribute their assignments as one complete project. Finally, the "think-pair-share" concept also illustrates the key role of group work and learning as a form of social interaction. Here, students are instructed to review a concept and write down their ideas. They are then instructed to share their ideas with one (or several) partners and compare their views with one another. The process of exchanging information helps to not only improve learning capacity but also builds the relationships among individuals within the group.

In each of these of cooperative activities, social interaction, communication and positive interdependency are vital components for successful learning. As Chomsky noted, language development was discontinuous in the sense that something dramatic must have evolved among human evolution, probably over 100,000 years ago. The existence of the protein FOXP2, first discovered among Neanderthals in the Denisova Caves of Siberia, Russia, suggests that humans at least had the capacity for formal language development and that cooperative learning was necessary in order for languages to be both taught and understood.

Humans have evolved with the very basic and primitive need to work together and to cooperate as a means of existence. The discovery of the FOXP2 gene suggests that humans began early language development approximately 100,000 years ago and that these primitive language skills were taught via cooperative mechanisms in small groups. The development of language itself among humans was dependent on our ability to work cooperatively and collaboratively within small group interaction. The capacity to share ideas through an expressed formal language greatly improved our ability to communicate among different clans and ultimately improved survival among each group. When groups of individuals work cooperatively with each other in the mastery of language development, their ability to resolve a broad range of social and economic as interpersonal problems is improved significantly (see Figure 8.1 below).

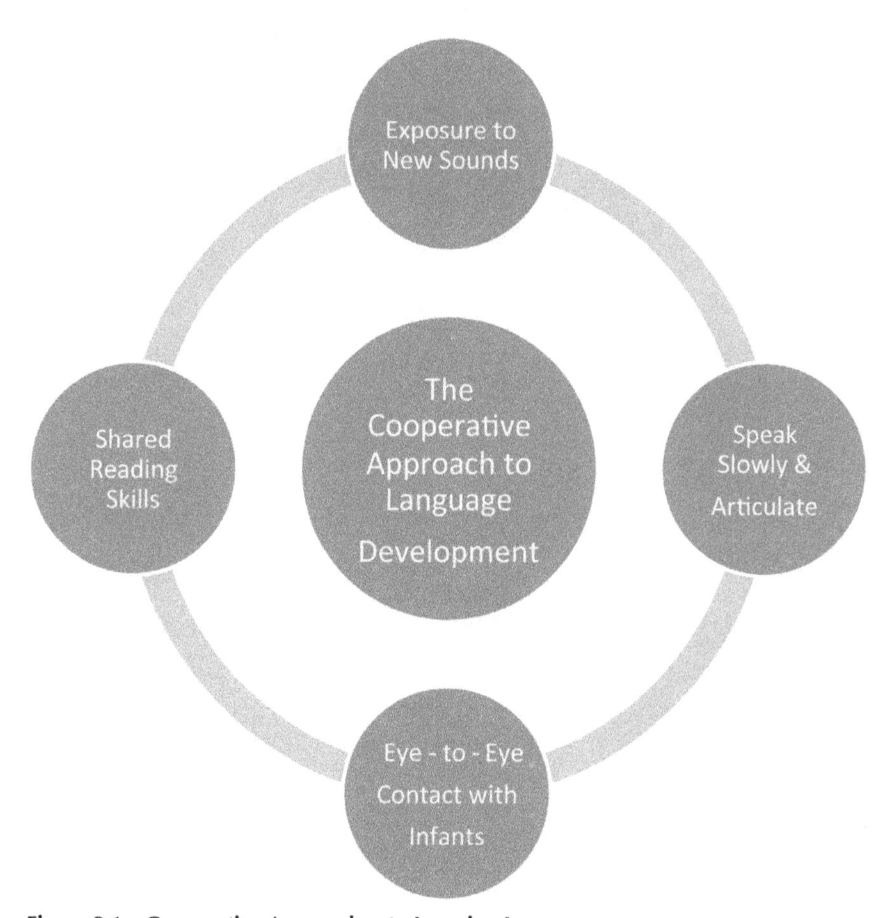

Figure 8.1 Cooperative Approaches to Learning Languages.

Part III

THE EVOLUTION OF MORALITY AND COOPERATIVE BEHAVIORS

Chapter 9

Teaching Virtue

How Cooperative Behaviors Improves Reproductive Fitness & Kinship

WHY DOES MORALITY EXIST IN HUMAN GROUPS?

Morality exists because without it humans would cease to exist. The ancient Greek philosophers Socrates and Plato argued that the single most important characteristic of humanity itself (and that which distinguishes us from other living things) is that of reason, morality and virtues. These things, as noble as they are, define our individual integrity in relation to others and are things not "learned" but rather discovered within ourselves through a process of self-discovery and introspection. According to the early Greek philosophers morality exists within each of us (as an innate phenomenon) and can only be manifested upon our own realization that we have the capacity to exercise these behaviors within groups and activities involving social interaction and community development.

Morality, then, serves as the monitor of human behaviors and evaluates the appropriateness of our own behavior in relation to other people within our group. Morality can often make us become far greater and noble individuals by placing concern of the welfare of others over our own well-being. True—greed and selfishness as characteristics of behaviors do exist today (some may argue increasingly so) but typically these traits exist only as a consequence of cultures that teach self-entitlement principles and encourage the development of "zero-sum" professional relationships. So . . . were the early Greek philosophers (i.e., Plato and Socrates) correct in assuming that the virtues of truth, honesty and justice may be realized among all persons within society? Clearly the answer is yes, but perhaps a more important question is in asking how we can restructure and design communities that will show each individual the value and importance of such behaviors so they are demonstrated among all groups of individuals.

Morality also exists because those individuals who do engage in a more selfless and altruistic fashion invariably improve the group's overall fitness and ability to withstand environmental and ecological pressures. Making small but meaningful sacrifices that emphasize sustainability to conserve fuel by walking to school or work instead of driving your car may not only help to conserve resources for future generations (i.e., non-kin members) but also reduce global warming. These kinds of socially and environmentally collective efforts not only help improve living conditions among our own kin but also help to improve overall conditions for other non-related individuals throughout the world.

However unique we may wish to appear in the development of altruistic and moralistic behaviors, animals of several varieties (i.e., wild dogs, bats, primates and even insects) have been recorded to engage in behaviors that results in their own demise but improves the colony's ability to withstand attacks from enemies. Humans, for example, may make a variety of personal sacrifices to families in need and participate in food donation programs for low-income families. While this is notable and admirable, it is by no means unique and even shared by similar animal behaviors. The common bat, for example, shares food to other members in the same colony who may not have been successful in securing food during the evening hunt.

Moral codes of behaviors that monitor and protect our relationships with others (even nongenetically related members) have been noted to increase inclusive fitness within those groups or clans. Simply stated, groups of humans or animals that make greater sacrifices for others within their own group have a significantly higher proportion of reproductive success in longevity of offspring, even if those offspring are from different mates within the same group and do not share same genetic lineage.

When discussing evolutionary biology, morality also is an important concept because sacrificial behaviors may exist to an individual but may improve the overall welfare of the group itself, and more specifically to one's own kin. The fact that humans can actually "feel" pity, shame or empathy has also been shown to be highly correlated with one's willingness to engage in sacrificial behaviors for others, or at least members of one's own group. Highly moralistic and ethical cultures throughout human civilization have encouraged ideological and philanthropic behaviors as a means of furthering cooperative behaviors. Small groups during evolutionary history that were more prone to share their skills (i.e., hunting, child-rearing or agriculture) emphasized the need for cooperative behaviors when sharing responsibilities that were vital for survival. Cooperative skills quickly formed during our early evolutionary history and helped those groups survive despite numerous threats to their health and safety.

An environment of *shared responsibility* helps to not only improve the overall inclusive fitness of the group itself, but perhaps more importantly can also improve the psychological states of well-being for each group member by establishing trust and interdependency within the group. The ability to work cooperatively with each group member provided the fabric of interconnectedness within the group and helped each member overcome conflict and friction during times of stress.

Research conducted by Bekoff and Pierce (2009) confirms this fact by noting that humans and large groups of animals typically share a broad range of skills that controls and regulates the living conditions within each group. The concept of "morality" therefore is not limited to humans, and actually several groups of animals share a variety of interrelated behaviors that serve as norms protecting the viability and reproductive success of each member within the group. Interestingly, Bekoff and Pierce (2009) define morality as: "a suite of interrelated behaviors shared by animals living in a complex social group."

Human prosocial behaviors were highly adaptive in the sense that they significantly increased and improved the welfare and reproductive fitness of the group itself. Despite the greedy and violent nature that Darwin's theory suggests, humans must have engaged in a variety of prosocial and cooperative behaviors that were characterized by a highly moralistic norm that governed the behaviors within the group. Thus, a highly collectivistic and group-oriented culture evolved that helped maintain the development and growth among humans in a highly cooperative society. The sacrifices that were made by some individuals within the group helped to maintain the overall safety and integrity of the group as it became larger and encountered more threats in the natural environment.

Related traits and characteristics of highly moralistic human behaviors have been noted by researchers Peterson and Seligman (2004). These researchers have noted near universal acceptance of specific traits or characteristics of moralistic behavior, such as virtue, honesty, temperance and wisdom. Historically these traits have served as a moral code that monitors individual behaviors and have helped establish a higher quality of living that protects all group members from harm and helps to prevent conflict by observing the rights of others within the group.

Children who are (vicariously) exposed to role models and group members engaging in a variety of these ideal and virtuous prosocial behaviors (i.e., honesty, temperance, and personal sacrifice) have been noted to be significantly more likely to engage in similar types of behaviors, including empathy. These effects have also been observed through the development of video games and media in general (Greitemeyer, Oswald, and Brauer 2010). Teaching younger populations the virtues of selfless behaviors over egoistic behaviors commonly portrayed in highly individualistic cultures can help groups overcome significant barriers and increase the quality of living for all members. When children participate in ethnically diverse community service work activities (CSW), they quickly learn the value of cooperation and collective self-efficacy. Similarly, when children are provided with opportunities to work with a diverse group of individuals they learn to understand different cultures and practices without criticism. They also show increases in empathy and reduced levels of self-entitlement attitudes because they have been provided with opportunities to see how different individuals from different cultural groups live and interact with each other (Hoffman, Wallach, and Sanchez 2010).

Exposure to group and community projects during developmental ages serves to teach children the value and importance of community service activities and increase the likelihood of future volunteer group projects. Conversely, children who may be exposed to the violence and opportunistic behaviors that are prevalent in society, often through violent media or video games, are more likely to emulate behaviors that historically have contracted the values and moral codes that have established and reinforced cooperative and prosocial behaviors within small groups and communities (Huesmann 2007; Greitemeyer, Oswald, and Brauer 2010). However, some recent research contradicts the evidence that suggests media influences behaviors among younger children (Ferguson 2013).

THE RELEVANCE OF EMPATHY, REVENGE AND FORGIVENESS IN COOPERATIVE BEHAVIORS

Early human development and human nature have often been depicted as egoistic, violent and opportunistic, where humans where described as savages primarily concerned only with their own survival (Freud, 1930: *Civilization and Its Discontent*). Human instincts and impulses were described as typically revengeful, sexual and violent, allowing little hope for group survival and civilization. The concepts of "forgiveness" and "kindness" do not readily come to mind when we discuss the nature of early human interaction. Somehow ideal human characteristics that promote prosocial behaviors were overlooked in the analysis of the human condition. Sigmund Freud depicted human nature within a reductionistic perspective, arguing that all humans are biologically and sexually destructive creatures who will ultimately cause their own apocalyptic destruction. Forgiveness, empathy and altruism were behaviors and traits that simply did not exist to many of the early psychologists and theorists when trying to understand how humans interacted with each other early on in human evolutionary history.

While it is true that vengeful and destructive behaviors (even murder) are universal characteristics of human interaction seen in over 60 different societies and cultures worldwide, it is also true that humans have evolved with the capacity to forgive even the most heinous crimes, such as blood revenge (McCullough 2008). McCullough argues that both revenge and forgiveness are universal human traits and that they have necessarily evolved to form a stronger cooperative human network. Forgiveness must have played an important role in monitoring group and individual behaviors during our evolutionary history given the numerous opportunities for transgressions to occur (i.e., mate poaching, theft, and conflict).

Mullet, Girard and Bakhshi (2004) have identified numerous positive psychological benefits to groups that display forgiveness, such as helping the transgressor to learn from past mistakes, reciprocal forgiveness (i.e., the former transgressor is now in a position to forgive others) as well as replacing negative emotions (i.e., vengeance) with positive emotions. More importantly, however, groups from collectivistic cultures that display qualities of forgiveness have also been shown to have stronger cohesiveness and solidarity (Fu, Watkins, and Hui 2004). Recent empirical research has shown numerous advantages in support of positive emotions, mental health and the capacity of forgiveness (Diener, Lucas, and Oishi 2002; Emmons and Paloutzian 2003). Individuals historically have engaged in a variety of different vengeful actions, dating back to Biblical scriptures ("You have heard the law that says the punishment must match the injury: 'An eye for an eye, and a tooth for a tooth.'" Matthew 5:38). The capacity to forgive also has been described as a primary instinct dating back to early human evolution (pre-Neanderthal) when humans lived and interacted in small groups or clans.

Trust has been described as an essential component in maintaining any social relationship that involves mechanisms of cooperation (McNight, Cummings, and Chervang 1998). Similarly, the capacity of forgiveness and empathy has also recently been identified as important features maintaining social harmony and cooperation during evolutionary periods of development (Balliet and Van Lange 2013). Groups that were competing for vital resources needed some degree of trust among each other as a means of increasing security and sharing those resources with each other. The concept of trust itself is important relative to evolutionary theory because it reflects awareness and belief of another group member's good intentions toward other group members (Yamagishi 2011). Invariably, during human evolution group members engaged in some form of conflict or transgression that tested the prosocial bonds of trust and ultimately required the capacity of forgiveness.

Until recently, the construct of forgiveness was typically avoided by social science researchers because of the perceived problems associated with complex psychological constructs, such as shame, spirituality and remorse. While forgiveness has been described in a multitude of ways involving different aspects of group and individuals situations, a summary of recent literature describes temporal change as one common denominator involving forgiveness (McCullough, Berry, Root Luna, Tabak, and Bono 2010). In a recent study exploring the functional relationship between forgiveness and time, McCullough and colleagues (2010) determined a logarithmic function between a negative stimulus (i.e., an argument of some form of transgression) and the passage of time based on the Weber-Fechner Law. Human evolution necessitated not only trust but also the capacity to forgive—small groups or clans in conflicts involving vital resources (e.g., food). McCullough and colleagues identified a key component involving the capacity to forgive in group dynamics: As time progressed proportionately so does our capacity to forgive in a linear fashion. The capacity to forgive over time represents a vital human characteristic in forging and maintaining positive relationships and social interaction. Harboring ill-will toward a transgressor with whom we typically have frequent contact with (i.e., holding a "grudge") not only deteriorates the integrity of positive future social interaction within that group but can also have a negative impact on physical health as well (Toussaint, Owen, and Cheadle 2012). In sum, as the capacity of forgiveness increases over time so does our capacity to trust others and develop stronger levels of cooperation when faced with adversity and conflict (Balliet and Van Lange 2013).

The capacity to forgive those who may have intentionally wronged us holds tremendous potential in our ability to form group solidarity and cohesiveness after some episode of in-group conflict. Franz de Waal (2005) argues that evolutionary theory supports the concept of forgiveness through the "valuable relationship" hypothesis. This hypothesis states that conflict is an inevitable characteristic of human behavior, and through the capacities of perspective taking and forgiveness we not only repair our relationships with other group members, but the overall cohesiveness and resiliency of the group becomes strengthened through a form of reciprocal reconciliation. In other words, those we have forgiven who have wronged us in the past are more likely to forgive us in the likely event that we engage in some form of a transgression against them in the future. So, in terms of evolutionary group dynamics, the process of forgiveness is similar to building a stronger network of relationships where people are more likely to trust you in future interactions.

McCullough argues that revenge evolved in human groups for several very important reasons. For example, revengeful action clearly prevented aggressors from attacking a specific group. If a group was known to protect itself through some type of revengeful action, a would-be perpetrator would be less inclined to make an initial attack. Knowing that a group will quickly retaliate if attacked or provoked serves as a strong incentive to respect the boundaries of other clans and groups. Another reason has to do more with social loafers or "free riders"—group members who are reluctant to participate in a variety of group responsibilities and activities (such as securing food). If a potential social loafer sees the negative consequences of violating a group norm (i.e., physical punishment), then he or she is more likely to conform to group responsibilities and help in future work assignments.

Several theories in this text have been presented that now dispute the claims that humans have evolved to be entirely aggressive and egoistic in their interactions with other groups. We introduce two new concepts relative to the development of cooperative behaviors: forgiveness and empathy. To "forgive" means to accept another's transgressions against us showing no malice or vengeance in future interactions with our transgressor. Empathy is also directly related to cooperative behaviors because it refers to the human capacity to understand and "feel" what others may be experiencing, such as "feeling your pain in the loss of your family member." Researchers know that it is easier (and more common) to forgive those who are closer to us—closer in both physical proximity (such as your neighbor) and emotional or spiritual closeness.

Inevitably, as humans evolved in clans and close living quarters in smaller communities conflicts would emerge within our own group as well as with "outside" groups. Realizing that without the aid, support and collective efforts of all group members our own survival would be jeopardized, early humans necessarily developed the capacity for forgiveness. Humans could not have existed in small groups without some form of empathy and forgiveness. In order for some type of prosocial and altruistic behavior to emerge, humans needed to develop the capacity to understand what others may have been experiencing.

Additionally, the capacity to forgive also required the ability to develop perspective taking of others and a recognition that in order for groups of individuals to live and interact in a productive and healthy manner, occasionally transgressions will develop that require us to acknowledge the mistakes of others and to move on in social development. Cultures that emphasize the adaptive value of cooperative relationships will see a higher proportion of forgiveness within the adaptive context of human interaction. A society that exists with less cooperative interaction will show a disproportionately higher rate of aggressive and revengeful activities. Thus, in order to reduce actions of aggression and revenge, cultures that emphasize group work and cooperative behaviors will be improving the quality of human interaction by reduced conflicts and increased opportunity for individuals to work things out in a collaborative manner.

Just as revenge and aggression have been documented in virtually all societies of the world throughout our evolutionary history, so does the capacity to forgive and to reconcile with others. Franz de Waal (2005) noted that the vast majority of primates often engage in highly aggressive behaviors, but interestingly in over half of observed aggressive actions (179 of 350 or 51%), the chimpanzees were observed to engage in what de Waal characterizes as "reconciliatory" behaviors: Cooing, touching, embracing and even kissing each other in what had appeared to the researchers as an effort to "make up" with each other, very similar to how humans show compassion after an altercation.

TIT-FOR-TAT STRATEGIES REVISITED: I FORGIVE YOU AND YOU FORGIVE ME

A key factor in group survival during early human evolutionary history was the development of social and group strategies that facilitated group survival. If individuals are rewarded for demonstrating a variety of positive and sacrificial behaviors (i.e., sharing food), they are more likely to engage in a variety of prosocial behaviors that collectively will benefit all members of the group. However, if one individual retaliates against other group members (i.e., defects from the group and engages in opportunistic behaviors such as stealing food from others), then the group typically will punish that individual.

Finally, accepting defectors in the aftermath of their transgression (i.e., stealing food) was vital in long-term evolutionary group development. The capacity to forgive our transgressors and bear no vengeance on him or her in future transactions typically is described as forgiveness. A common theme developed early on in human interaction that basically argued for a "Tit-for-Tat" principle: If others treat you well in your group, then reciprocate. However, if others defect and try to steal from you, punish the individual as a group and then forgive. Clearly, without the evolution of the virtue of forgiveness, human survival within groups and small clans would have been impossible. The capacity of forgiveness has also been shown to be a significant trait that helped to reduce intergroup conflict and promote a stronger moral identity to within group members (Reed and Aquino 2003).

To cooperate in any group activity means to contribute to the development of some end product or goal—whether that means cooperating to catch prey for the survival of your group members or working together to build a home to live. In order for humans to cooperate, we need to be aware of what others are doing and how we may contribute. Empathy plays a vital role in cooperation because it helps us to see and to personally experience how others feel in the development of a skill or tool that is useful to the integrity and strength of the group itself.

ARE HUMANS UNIQUE IN THE DISPLAY OF EMPATHY?

Evolutionary psychologists and anthropologists have noted that those females who responded to the biological needs of their offspring fastest tended to have those offspring more resilient and survive when compared with mothers with a slower response time to the needs of their offspring (De Waal 2005). Part of the reason why Darwin's theory of evolution has created the "shock and awe" in the general public during the Victorian Era was the (incorrect) assumption of the distinct intellect of humans. People had assumed that humans alone possessed formal language, superior intellect and prosocial behaviors, such as altruism when compared to other animals. We have identified throughout this monograph that those assumptions are invalid—primates have a well-established communication system (even sign language, as evidenced by Koko, the gorilla), bats and rodents have been observed to frequently share their resources such as food with lame or disabled group members, and in some cases demonstrate altruism by sacrificing its life for the safety and integrity of the herd. One more characteristic that humans and animals share in group development and cooperative behaviors is that of empathy.

Empathy has been described in several ways, but generally refers to the capacity for one individual to share the perspectives, feelings and emotions with other animals. Young chimps have been observed to actually refuse to activate devices that are known to cause pain (i.e., electric shocks) to other chimps (American Journal of Psychiatry 1964) and anthropoid apes express distress at the sight of other apes suffering from illness (Yerkes 1925). In some situations involving conflict, chimps have been observed to engage in consoling behaviors such as embracing each other and in some situations kissing each other, in an apparent effort to achieve reconciliation (McCollough 2008). Both of these examples serve to illustrate how animal groups from diverse species can understand when other animals are experiencing some form of distress and discomfort.

Primatologists note with consistency that primates (i.e., chimps, gorillas and orangutans) share the human skill of "perspective taking" from cross-species. Koko, the famous gorilla of Penny Patterson, mourned the death of her kitten and Kuni, a bonobo female in Twycross Zoo (England) helped recuperate an injured bird while living in her outdoor cage. Lower primates and early humans not only developed the capacity of empathy and forgiveness to serve the needs of other group members, but they also developed the skill to create and forge a stronger and more resilient group during evolutionary development. Empathy plays a critical role in cooperative behaviors as we need to understand what others are actually feeling in order to motivate us to help in some way.

Franz de Waal (2005) notes that a lioness needs to actually see other lions hunting in a group for prey before she decides to join the pride herself. Empathy refers to our capacity to experience what we think others may be experiencing as well assume the perspectives of others (kin or stranger). When we see and understand the needs of others we generally become motivated to help in some way. Empathy allows us to "feel" what others feel and this capacity serves as both incentive and motivation to help others in distress. This function not only served to promote individual relationships but also helped us to secure a stronger and resilient group in times of danger and stress.

The psychological traits of empathy and perspective-taking help us to better understand from an evolutionary perspective what others may be thinking and thereby reduce potential conflict through the capacity of forgiveness. An important incentive of forgiveness is in trying to understand what other people may have been thinking and feeling prior to their transgression toward us—in other words—getting into the heads of others. Logically, then, it is easier to forgive when we can see and understand what led someone to try to harm us in some way. Jealousy and envy are two common factors that play a role in intergroup conflict. We may be initially put off by frequent insults and critical remarks by a coworker or neighbor, but when we realize that the criticisms are essentially due to something that they actually admire in us, we often change how we feel about that person, and in some cases may even be flattered.

In addition to the psychological traits of empathy and perspective taking as effective measures of instilling forgiveness is that of group contact. The topic of group contact has been studied extensively over the years as an effective method and procedure in improving group relationships and removing potential conflict from intergroup behaviors (Allport 1954; Pettigrew 1998). While much research has been conducted addressing how contact within group development helps to reduce conflict and related problems such as prejudice and ethnocentrism (Pettigrew and Tropp 2006), we argue that group contact is really only addressing half of the problem in trying to reduce social problems such as intergroup conflict and prejudice. In order for contact to actually help us understand and actually engage with others within our group, we need to first somehow understand that the negative stereotypes previously held about that person or group are invalid and we need to determine that individuals initially perceived negatively are actually more similar to ourselves.

Contact alone does not make us better neighbors or more sympathetic to the plights of others; in order to significantly reduce conflict and prejudice group members need to establish an interdependent relationship with others and work collectively in trying to achieve a common goal that is valuable to all group members. Group work that requires *active involvement* and participation in a mutually satisfying end product or result combined with applying varied skills from each group member is the most effective process in reducing intergroup conflict as well as prejudice and ethnocentric ideologies (Hoffman, Wallach, and Sanchez 2010).

When individuals are provided with a multitude of different opportunities to share their skills with each other, group conflict and polarization are typically reduced. These common social problems are reduced because typically negative stereotypes addressing other groups (ethnicity, gender, etc.) have been debunked and groups can focus on individual strengths that each member invariably has. Individuals work collaboratively with each other to create a stronger and more resilient community and society. Similar to our evolutionary ancestors, the need to contribute to the wellness and development of our society exists in each of us because of the universal need to survive—traits that exist in our genetic fiber and DNA.

It is important to note here that this does not mean that all individuals within our society actually engage in the cooperative behaviors that build society, but we all have the genetic predisposition to do so. Simply because an individual has the genetic skills and aptitudes to be an athlete, artist or automechanic does not mean that *ipso facto* it will happen—what actually defines the outcomes of a specific human behavior is a very complex social and environmental process. Environmental, social and ecological factors actually determine if these behaviors will be manifested throughout human development. When communities fail to provide individuals with opportunities to work collaboratively and to share in the process of learning and contributing their skills with others within a group setting, the groups then become disjointed and polarize with perceived phenotypical similarities which are counterproductive to interdependent and supportive group relationships.

In the beginning of this monograph the tragic situations of the Sandy Hook Elementary School, Columbine High School and the Emanuel African Methodist Episcopal Church in South Carolina were used as several examples how alienation from the community and group development can potentially contribute to hate crimes and psychopathological behaviors. These national tragedies were committed by individuals (Adam Lanza, Eric Harris, Dylan Klebold and Dylann Roof) who many would describe as "loners"—people who had (initially) attempted to connect with their community but for several reasons were unable to do so. In each case, the individuals responsible for the shootings were preoccupied with guns, weapons of mass destruction, violent media and antisocial video games (i.e., *Doom* and *Call of Duty*). A recent *New York Times* article ("*Newtown Killer's Obsession, in Chilling Detail*" March 28, 2013) describes much of Adam Lanza's behaviors prior to the Newtown, CT shooting as withdrawn, engaged in isolated activities and playing video games such as *Call of Duty*.

Finally, it is very important to note that these accounts do not support the claim that the violent media and video games were actual *causal* factors of the shootings, as Ferguson (2013) notes. To make the claim that one variable, such as violent media games, was the sole contributor to highly antisocial behaviors would be incorrect. However, the isolated and polarized behaviors of the individuals responsible for the tragic shootings were associated with a lack of community involvement and interpersonal connectedness with friends and community members. The primary mission in writing this monograph has been in identifying the process in which early Greek philosophers attempted to define the relationship between the mind and the body and how this philosophy influenced the development of modern psychology and evolutionary psychology. Evolutionary psychology has shown us that a better understanding of what was necessary in surviving throughout human evolution is still necessary today—this would include a supportive and interactive community that encourages interdependency where individuals feel connected to each other and share a sense of self-worth. Through millions of years of evolution humans have developed both the capacity and need to live cooperatively and socially with others in their environment. When we become displaced in an environment that prevents this form of cohesion and shared communication, problems and conflicts will inevitably develop.

In conclusion, communities not only have the capacity but also more importantly an obligation to create specific environments that stimulates and nurture cooperative relationships, such as the development of community service activities and civic engagement opportunities. The early Greek philosophers such as Plato and Socrates remind us of the inherent good that lies within each of us—temperance, virtue, introspection and honesty are just some of the characteristics that need to be discovered (through our own self-analysis) within our interactions with others within the community. Communities that provide opportunities for growth, development and social capital are strongest because each individual views himself/herself as a vital and contributing component within a larger social system.

In a word, all lives do matter through our innate capacity to recognize the skills that each person possesses and is capable of sharing within the community. Sharing our skills is the most effective process in understanding how we may best "fit" within the community and understand people from diverse backgrounds. Providing community members with opportunities to work collectively and share skills with each other helps everyone understand and identify our common needs as well as our common bonds. A morally articulate and empathic society that understands the needs of all persons will also help individuals to develop an important sense of "connectedness" to their own community. Conversely, when communities fail to provide these critical opportunities of human interaction, growth and development we will continue to see tragic events unfold similar to those in Newtown, CT, Boulder, CO and Charleston, SC.

Epilogue

Notes for the Future: Future Development of Evolutionary Psychology

Although humans (*Homo erectus*) have existed for literally millions of years, we were not the only species to have existed on this planet. Other species of humans (or human-like creatures), referred to as hominids, have existed for shorter periods of time but for some reason became extinct within the last several million years. For example, whereas Neanderthals existed as recently as 40,000 years ago in Western Europe (southern France), a shorter, less muscular species (*Homo floresinsis*) existed approximately 60,000–80,000 years earlier. Before *Homo floresinsis*, the species *Paranthropis robustus* existed, and before then, *Australopithicus sediba*. Evolutionary psychologists and anthropologists have identified over a dozen different types of apes and hominids that have lived for shorter periods of time, but only one species, *Homo sapiens*, have survived the true test of time.

What were the distinguishing factors that enabled *Homo sapiens* or modern humans to continue their existence, despite conflict, war, disease and famine? Prosocial behaviors, altruism or the capacity of forgiveness and empathy? While many (animal) predators may have contributed to the demise of our early ancestors, the greatest contributing factor to human mortality is through yet another human being (Suddendorf 2013). This shocking fact remains true even today. And yet an interesting paradox emerges when considering the politics of human survival—our ability to survive increasingly depends upon how well we can share our own resources and resolve potential conflicts with other groups. Human nature has a simultaneous capacity or duality of both inherent strengths for survival and growth as well as violence and destruction among one another. In a word, humans needed to learn to understand and resolve their differences in order to create a safer and more productive environment for the safety of all community and group members.

Human survival (or death) hinges on the development and proliferation of human cooperation, communication and the sharing of resources. Similar to the Prisoner's Dilemma, the best strategy is one that supports a mutually cooperative approach where we protect our own interests by protecting our partner's interests as well often referred to as the "tit-for-tat" approach. Without human cooperation, aggression and conflict may have been inevitable in the constant battle for survival over scarce resources (i.e., food and fuel for building fire) as well as reproductive resources such as women of child-bearing ages.

Throughout this monograph both philosophical and psychological arguments have been offered for the evolutionary development of an essential behavior to human growth and development: Cooperation. To cooperate with others requires an ability to recognize the needs of others and relate to how others may feel during times of need and crisis. Ironically, during our evolutionary history those periods of great stress and anxiety often resulted in people responding with the greatest compassion, altruism and empathy. The capacity of cooperation also necessitates the ability for us to (temporarily) set aside our own needs and problems and to focus on the needs of others. The importance of this trait was addressed centuries earlier through the works of Socrates and the Socratic Paradox: We stand to gain much by assuming that we are capable of knowing very little, and in a similar way we help ourselves when we help others. This admirable trait is perhaps one of the defining and distinctive characteristics that separate the traditional collectivistic (group-oriented) culture from the more competitive individualistic cultures those we so often read about in sports and entertainment news clips.

SOCIAL MEDIA, SELF-ENTITLEMENT AND INDIVIDUALISTIC IDEOLOGY: KEY THREATS TO COOPERATIVE BEHAVIORS

Perhaps the greatest threat to cooperative behaviors and community-based activities are self-interest and self-entitlement attitudes that have been exacerbated by the significant increases in social media. It is difficult to speculate how Plato or Socrates would have responded to social media during their lifetime. Perhaps, if these tools were used wisely in the capacity to increase our knowledge about ourselves and improve the context of our relationships with others, then the early philosophers may have actually welcomed the arrival of social media and electronic technology. The concept of self-entitlement refers to an expectation of others to do what we should more appropriately be doing for ourselves. Self-entitlement typically is reflected in what individuals expect others to do to improve social and environmental conditions around them.

Many theorists discussed previously in this text have noted that human nature has a tendency to engage in what many would consider egoistic and selfish behaviors and that the only justification for accessing anything and everything is merely the capacity to do so. Thomas Hobbes, for example, argues that human nature itself is prone to destruction and violence in the desire to acquire all things pertaining to value. In Leviathan, Hobbes argues for a Social Contract that allows all community members to share their skills and contribute to commerce without fear of anarchy. To make this possible—some unifying and central force, such as governments, are necessary to protect communities and the civic interests of individuals living in society. What Hobbes was basically asking for in very simple language is the freedom for communities to organize themselves into labor forces under a loosely defined governmental agency that protects all individuals from the greed and destructive capacities that are also inherent in all individuals.

Unfortunately, the words of Thomas Hobbes in the sixteenth-century still ring true today when describing human nature. Civic engagement, volunteerism and community service activities are all examples of behaviors with one common underlying theme: Cooperation. Cooperative behaviors are essential in the development of activities involving individuals within a variety of community-related projects as they provide a forum that allows individuals to share and contribute their skills to other community members. Humans and even earlier human ancestors (hominids) evolved with the capacity to share skills and resources in the never-ending battle to survive. Thus, human cooperation is more than just a choice; human cooperation is a genetically evolved trait that provides the capacity for individuals with the skills to improve their ability to survive with each other (Sterelny, Joyce, Calcott, and Fraser 2013).

There have been several positive developments noted within the last several years in the development of social media, such as increased communication with previous "dormant" relationships, maintaining close social ties and even reports that individuals (i.e., older populations) are becoming more politically engaged (Pew Research Center 2011). However, despite these advances individuals still report feeling more "disconnected" with each other than ever before. Related problems such as cyber-bulling, decreased levels of productivity within the workplace, as well as increased false perceptions of connectedness within our social and narcissism world have all been exacerbated through social media (Carpenter 2012; Greenwood 2013). Related to some of the problems associated with social media have been distractions in academic work and productivity (Junco 2012). Many of the social problems that have been associated with social media exist because individuals are increasingly spending more time with technology and less time engaged in interpersonal activities that are designed to help improve communities and social well-being.

Community service work opportunities continue to provide individuals with unique opportunities to work and collaborate with each other in real time. The needs for environments that promote cooperative and interactive behaviors are still very much an important component of human existence that can help reduce conflict and aggression.

HUMAN EVOLUTION, GROUP WORK AND INTERDEPENDENCY

People have evolved with the basic need to contribute and share their unique skills and traits with each other within a community. Sharing these skills provide stronger bonds with each individual and make the group or community both more interactive and resilient. Humans have long needed opportunities to work, socialize and interact with each other as a means of building trust and support (and in some cases altruistic exchanges) within our relationships with other community members. Throughout our evolutionary history, humans have needed to work together, collaborate and share their skills and resources in an effort not only to survive but also as a means of thriving in very harsh conditions (Buss 2012). A collaborative work effort also provides opportunities for community members to share their skills in the effort of strengthening and building their towns and communities.

Communities that provide individuals with opportunities to work collaboratively with each other (i.e., community service work activities) and share their skills have reported a stronger sense of "community connectedness" as well as a sense of "psychological sense of community" (Putnam 2000). Communities that share community service work opportunities also experienced a stronger sense of connectedness and improved relationships with diverse community members based on those activities (Hoffman, Wallach, and Sanchez 2010). More recently, civic engagement practices and involvement with urban forestry programs (i.e., planting trees in densely populated cities) have been identified as key mechanisms in the development of psychological healing and establishing disaster relief to communities in the wake of tragedy (Tidball and colleagues 2010)

As communities become more advanced in the multiple uses of social media and electronic technology as a means of interacting and communicating within their social and professional lives, the opportunity for direct (i.e., nonviral) interpersonal community service work has rapidly declined (Putnam 2000). More often individuals report preferring to engage with and "hook up" with friends via social media (i.e., Facebook) and rely on a variety of technologically related devices as a means of social interaction (Rieber and Garcia 2010). Direct and interpersonal community service work and volunteer activities have recently become less popular and fewer communities are now offering opportunities for community members to volunteer and participate in community service projects (Hoffman, Wallach, Espinoza Parker, and Sanchez 2009).

The psychological and social advantages to communities that provide community service work opportunities to residents are numerous and have been supported through empirical research (O'Connor and Jose 2012). Conversely, as community service work activities tend to decrease over the years for a variety of reasons, interest in video games and technology is rapidly increasing in societies throughout the United States, where almost 70% of households in the United States are currently using some form of technology-related video game device. Fewer community service work activities are now being made available for community members which is now having a negative impact on the capacity for holistic community growth, development and connectedness (Putnam 2000).

Providing opportunities for community members to participate in community gardening programs can help low-income families provide healthier foods and establish a healthier standard of living (Carney, Hamada, Rdensinski, Sprager, Nichols, Liu, Pelayo, Sanchez, and Shannon 2012) as well as providing members from different ethnic groups to work collaboratively with each other (Hoffman, Wallach, and Sanchez 2010).

Communities today have unfortunately experienced fewer opportunities for participation in civic engagement and community service work activities for a variety of reasons: Changing economic structures within society, differences in how technology creates and shapes relationships through social media, and a lingering and persuasive individualistic ideology that encourages an emphasis on the needs of the self that is paramount to the needs of the community and concerns of society. Collectivistic cultures instill values that are central to the model of evolutionary past history and remain tantamount to what community service work activities represent to the welfare of our society. Teaching younger individuals in primary and secondary institutions of educations the values and benefits of helping others via community service work and stewardship offers several important benefits.

When individuals are exposed to various models of civic engagement and community service work activities, they are afforded the opportunities to experience first-hand positive reform in the communities in which they interact and reside in. Community service work activities also provide unique experiences to work with individuals from diverse economic, religious and ethnic backgrounds that have been empirically validated to reduce common problems associated with multiethnic environments (i.e., ethnocentrism, racism and sexism). These negative stereotypes have been reduced (and in some cases eliminated altogether) because community service work activities provide opportunities for individuals to see that we as a community and as a human race have much more in common than our perceived differences.

CONCLUDING NOTES REGARDING COOPERATIVE BEHAVIORS

It would be presumptuous to speculate that a stronger emphasis in community service and cooperative programs such as the community gardening programs described throughout this chapter would have prevented the highly destructive and horrific events at the Sandy Hook Elementary School, Columbine High School and the Emanuel African Methodist Episcopal Church in Charleston, South Carolina. However, if we as a community wish to make some type of proactive effort in trying to stop these horrific events, we need to learn to identify the "red flag" warning signs that we so often hear after the tragic event has occurred. Individuals who show a preoccupation with violent media that result in a failure to establish meaningful interpersonal relationships with others is a common theme that has been associated with highly destructive and violent behavior.

Perhaps the single greatest advantage to community involvement and community service work activities is the highly positive reciprocal relationship that we all experience when we are capable of "giving back" to our community. Simply knowing how to connect with others and communicate with other community members can help us to identify with and support other individuals within our societies, school districts and community environments. Similar to the "Socratic Paradox": We gain much more in the sense of community connectedness and belonging only when we first make personal sacrifices to improve our communities and we are capable of learning more by readily acknowledging our own ignorance. Understanding the basic evolutionary and biological dynamics of group interaction is critical not only for community development but for our very survival as a species.

"How to Teach the Six Basic Principles of Cooperative Behaviors"

- ✓ Get Out of Your Comfort Zone. Change is good! Be ready to make personal sacrifices (Turn off televisions, stop texting) and start experiencing events and making positive changes in your community;
- ✓ Get Involved in Your Community: what kinds of changes would you like to see? Safer communities, better school systems and cleaner

environments require individuals to work collaboratively and become involved in issues that benefit everyone;

- ✓ Teach Basic Human Values: Involve younger people in a variety of community projects to see and to experience other person's experiences—Compassion, empathy and stewardship are all "teachable" entities that children need to be exposed to. The capacity of empathy means experiencing what other's experience;
- ✓ Reduce Self-Entitlement Attitudes: It's Not All About You! Expect less and be willing to do more for your community. Listen to others who live in different cultures, be willing to change your belief systems that maintain the "status quo";
- ✓ Be a Role Model: Remember—those individuals watching you (children, students, other community members) are most likely to imitate the kinds of things and activities that you are involved with—what kind of role model are you?
- ✓ Respect Others & Respect Environment: Be willing to listen to others and respect their views event though they may be very different from your own.

References

Ainsworth, S. E. & Maner, J. K. (2012). Sex begets violence: Mating motives, social-dominance, and physical aggression in men. *Journal of Personality and Social Psychology, 103*, 819–829.

Allport, G. W. (1954). *The Nature of Prejudice*. Cambridge, MA: Additson-Wesley.

Anderson, C. A., Berkowitz, L., Donnerstein, E., Huesmann, L. R., Johnson, J., & Linz, D. (2003). The influence of media violence on youth. *Psychological Science in the Public Interest, 4*, 81–110.

Archer, J. (2004). Sex differences in aggression in real-world settings: A meta-analytic review. *Review of General Psychology, 8*, 291–322.

Arciniega, G. M., Anderson, T. C., Tovar-Blank, Z. G., & Tracey, J. G. (2008). Toward a fuller conception of Machismo: Development of a traditional Machismo and Caballerismo scale. *Journal of Counseling Psychology, 55*, 19–33.

Arnon, I. & Ramscar, M. (2012). Granularity and the acquisition of grammatical gender: How order of acquisition affects what gets learned. *Cognition, 122*, 292–305.

Aronoff, J., Wioke, B.A., & Hyman, L. M. (1992). Which are the stimuli in facial displays of anger and happiness? Configurational bases of emotion recognition. *Journal of Personality and Social Psychology, 62*, 1050–1066.

Arsenberg, B., Schepartz, L. A., Tillier, A.M., Vandermeersch, B., & Rak, Y. (1990). A reappraisal of the anatomical basis for speech in middle Paleolithic hominids. *Journal of Physiological Anthropology, 83*, 137–146.

Atran, S. (1998). Folk biology and the anthropology of science: Cognitive universals and cultural particulars. *Behavioral and Brain Sciences, 21*, 547–609.

Axelrod, R. (1984). *The Evolution of Cooperation*. New York, NY: Perseus Books Group.

Badcock, P. B. (2012). Evolutionary systems theory: A unifying meta-theory of psychological science. *American Psychological Association, 16*, 10–23.

Baldwin, J. M. (1896). A new faction in evolution. *American Naturalist, 30*, 441–451.

Balliet, D., Macfarlan, S. J., Li, N. P. & Van Vugt, M. (2011). Sex differences in cooperation: A meta-analytic review of social dilemmas. *Psychological Bulletin, 137*, 88–909.

Balliet, D. & Van Lange, P. (2013). Trust, conflict and cooperation: A meta-analysis. *Psychological Bulletin, 139*, 1090–1112.

Bandura, A. (1977). *Social Learning Theory*. Englewood Cliffs, NH: Prentice Hall.

Bandura, A. (1986). *Social Foundations of Thought and Action*. Englewood Cliffs, NJ: Prentice-Hall.

Bandura, A., Fernandez-Ballesteros, R., Diez-Nicolas, J., Caprara, G. V., & Barbaranelli, C. (2002). Determinants and structural relation of personal efficacy to collective efficacy. *Applied Psychology: An International Review, 51(1)*, 107–125.

Baron, R., Branscombe, N. R., & Byrne, D. (2008). *Social Psychology* (12th ed.). Pearson Publishers.

Barrett, H. C. & Kurzban, R. (2006). Modularity in cognition: Framing the debate. *Psychological Review, 113*, 628–647.

Bartlett, T. (2012). A New Twist in the Sad Saga of Little Albert. The Chronicle of Higher Education, http://chronicle.com/blogs/percolator/a-new-twist-in-the-sad-saga-of-little-albert/28423

Bates, E. (1976). Language and context: The acquisition of pragmatics. In: B. MacWhinney (Ed.), *Mechanics of Language Acquisition*, Hillsdale, NJ: Erlbaum publishers.

Baumeister, R. F. & Sommer, K. L. (1997). What do men want? Gender differences and two spheres of belongingness: Comment on Cross and Madson. *Psychology Bulletin, 122*, 38–44.

Bekoff, M. & Pierce, J. (2009). *Wild Justice: The Moral Lives of Animals*. Chicago, IL: The University of Chicago Press.

Blossner, N. (2007). The city-soul analogy. In: G. R. F. Ferrari (Ed.), *The Cambridge Companion to Plato's Republic*. Cambridge: Cambridge University Press.

Blumenthal, A. (1980). Wilhelm Wundt and early American psychology. In: R. Rieber (Ed.), *Wilhelm Wundt and the Making of a Scientific Psychology* (pp. 117–135). New York, NY: Plenum Press.

Blurton-Jones, N. G. (1972). Non-verbal communication in children. In: R. A., Hinde (Ed.), *Non-verbal Communication*. New York, NY: Cambridge University Press.

Bobo, L. (1983). Whites' opposition to busing: Symbolic racism or realistic group conflict? *Journal of Personality and Social Psychology, 45*, 1196–1210.

Boone, R. T. & Buck, R. (2003). Emotional expressivity and trustworthiness: The role of nonverbal behavior in the evolution of cooperation. *Journal of Nonverbal Behavior, 27*, 163–182.

Borkenau, P. & Liebler, A. (1992). Trait inferences: Sources of validity at zero acquaintance. *Journal of Personality and Social Psychology, 62*, 645–657.

Bowles, S. (2006). Group competition, reproductive leveling, and evolution of human altruism. *Science, 314*, 1569–1572.

Bradbury, J. W. & Vehrencamp, S. L. (1998). *Principles of Animal Communication*. Sunderland, MA: Sinauer Associates, Inc.

Buck, R. (1975). Nonverbal communication of affect in children. *Journal of Personality and Social Psychology, 3*, 644–653.

Burt, J. W. (1995). Dissertation Abstracts International: *Section B: The Sciences and Engineering*, Vol. 56(5-B), Nov, 1995, p. 2895.

Buss, D. M. (1989). Sex differences in human mate preferences: Evolutionary Hypothesis tested in 37 different cultures. *Behavioral and Brain Sciences, 12*, 1–49.

Buss, D. M., Larsen, R. J., Westen, D. & Semmelroth, J. (1992). Sex differences in jealousy: Evolution, physiology and psychology. *Psychological Science, 3(4)*, 251–255.

Buss, D. M. (2005). *The Handbook of Evolutionary Psychology*. New Jersey, Hoboken: John Wiley & Sons, Inc.

Buss, D. M. (2008). *Evolutionary Psychology: The New Science of the Mind* (2nd ed.). Boston: Allyn and Bacon.

Buss, D. M. (2009). The great struggles of life: Darwin and the emergence of evolutionary psychology. *American Psychologist, 64*, 140–148.

Buss, D. M. (2015). *Evolutionary Psychology: The New Science of the Mind*. Boston, MA: Pearson Publishers.

Buss, D. M. (2012). Pathways to individuality: *Evolution and the development of personality traits*. Washington, DC: American Psychological Association.

Buss, D. & Shackelford, T. (1997). From vigilance to violence: Mate retention tactics in married couples. *Journal of Personality and Social Psychology, 72*, 346–361.

Buss, D. & Reeve, H. K. (2003). Evolutionary psychology and developmental dynamics: Comment on Likliter and Honeycutt. *Psychological Bulletin, 129(6)*, 848–853.

Buss, D. & Shackelford, T. K. (1997b). Human aggression in evolutionary psychological perspective. *Clinical Psychology Review, 17*, 605–619.

Carney, P. A., Hamada, J. L., Rdesinski, R., Sprager, L., Nichols, K. R., Liu, B. Y., Pelayo, J., Sanchez, M. A., & Shannon, J. (2012). Impact of a community gardening project on vegetable intake, food security, and family relationships: A community-based participatory research study. *Journal of Community Health, 37*, 874–881.

Carpenter, C. J. (2012). Narcissism on Facebook: Self-promotional and anti-social behavior. *Personality and Individual Differences, 52*, 482–486.

Cartwright, J. (2000). *Evolution and Human Behavior*. Cambridge, MA: MIT Press.

Carvello, M. & Pelham, B. W. (2006). When fiends become friends: The need to belong and perceptions of personal and group discrimination. *Journal of Personality and Social Psychology, 90,* 94–108.

Cashdan, E. (1998). Are men more competitive than women? *British Journal of Social Psychology, 37*, 213–229.

Cattaneo, L. B. & Goodman, L. A. (2015). What is empowerment anyway? A model for domestic violence practice, research and evaluation. *Psychology of Violence, 5*, 84–94.

Charles, E. P. (2013). Psychology: The empirical study of epistemology and phenomenology. *Review of General Psychology, 17*, 140–144.

Chomsky, N. (1957). *Syntactic structures*. The Hague: Mouton.

Christiansen, M. & Kirby, S. (2003). *Language Evolution*. New York, NY: Oxford University Press.

Clutton-Brock, T. H. & Parker, G. A. (1992). Potential reproductive rates and the operation of sexual selection. *Quarterly Review of Biology, 67*, 437–456.

Cole, J. (1997). *About Face*. Cambridge: MIT Press (A Bradford Book).

Confer, J. Easton, J., Fleischman, D., Goetz, C., Lewis, D., Perilloux, C. & Buss, D. (2010). Evolutionary psychology: Controversies, questions, prospects, and limitations. *American Psychologist, 65(2)*, 110–126.

Cook, M., Minka, S., Wolkenstein, B., & Laitsche, K. (1985). Observational conditioning of snake fear in unrelated rhesus monkeys. *Journal of Abnormal Psychology, 93*, 355–372.

Coppens, P. (2012). http://philipcoppens.com/socrates.html

Cosmides, L., Tooby, J., & Barkow, J. (Eds.). (1992). Evolutionary psychology and conceptual integration. *The Adapted Mind* (pp. 3–18). New York, NY: Oxford University Press.

Creel, S. & Creel, N. M. (2002). *The African Wild Dog: Behavior, Ecology and Conservation*. Princeton, NJ: Princeton University Press.

Cross, C. P., Copping, L. T., & Campbell, A. (2012). Sex differences in impulsivity: A meta-analysis. *Psychological Bulletin, 137*, 97–130.

Curhan, K. B., Sims, T., Markus, H. R., Kitayama, S., Karasawa, M., Kawakami, N., Love, G. D., Coe, C. I., Miyamoto, Y., & Ryff, C. D. (2014). Just how bad negative affect is for your health depends on culture. *Psychological Science, 25*, 2277–2280.

Curran, E. (2002). Hobbes' theory of rights: A modern interest theory. *The Journal of Ethics, 6(1)*, 63–86.

Curtiss, S. (1977). Genie: A psycholinguistic study of a modern-day "wild child." *Perspectives in Neurolinguistics and Psycholinguistic*s, Boston, MA: Academic Press.

Daly, M. & Wilson, M. (1983). *Sex, Evolution, and Behavior* (2nd ed.). Boston: Willard Grant Press.

Daly, M. & Wilson, M. (1988). *Homicide*. Hawthorne, NY: Aldine.

Darwin, C. (1859). *The Origin of Species*. London: John Murray.

Darwin, C. (1887). *The Life and Letters of Charles Darwin, Including an Autobiographical Chapter*. London: John Murray (The Autobiography of Charles Darwin).

Darwin, C. (1871). *The Descent of Man, and Selection in Relation to Sex*. London: John Murray.

Darwin, C. (1872). *The Expression of the Emotions in Man and Animals*. London: Murray.

Darwin, C. (1959). *On the Origin of Species by Means of Natural Selection, or, the Preservation of Favoured Races in the Struggle for Life*. London: J. Murray.

Dawkins, R. (1976). *The selfish Gene*. New York, NY: Oxford University Press.

De Herdt, T. (2003). The Flood-Dresher experiment revisited. *Review of Social Economy, 41*, 183–210.

Denollet, J., Martens, E., Nyklicek, I., Conraads, V., & de Gelder, B. (2008). Clinical events in coronary patients who report low distress: Adverse effect of repressive coping. *Health Psychology, 27(3)*, 302–308.

DePaulo, B.M., Lindsay, J.J., Malone, B.E., Muhlenbruck, L., Chandler, K., & Cooper, H. (2003). Cues to deception. *Psychological Bulletin, 129*, 74–118.

Desmond, J. (1997). *Huxley: From devil's disciple to evolution's high priest*. Reading, MA: Addison-Wesley.

DeWaal, F. B. M. (2005). *Our Inner Ape: A Leading Primatologist Explains Why We Are Who We Are*. Penguin Books Ltd., Riverhead Books.

Dewsberry, D. A. (2009). Charles Darwin and psychology at the Bicentennial and Sesquicentennial: An Introduction. *American Psychologist, 64*, 67–74.

Diener, E., Lucas, R., & Oishi, S. (2002). Subjective well-being: The science of happiness and life satisfaction. In: C. R. Snyder & S. J. Lopez (Eds.), *Handbook of positive psychology* (pp. 463–473). London, Oxford University Press.

Dollard, J., Doob, L., Miller, N., Mowerer, O. H., & Sears, R. R. (1939). *Frustration and aggression*. New Haven, CT: Yale University Press.

Donnerstein, E. (2011). The media and aggression: From TV to the Internet. In: J. P. Forgas & A. W. Kruglanski (Eds.), *The Psychology of Social Conflict and Aggression* (pp. 267–284). New York, NY: Psychology Press.

Dorter, K. (2006). *The Transformation of Plato's Republic*. Lexington Books.

Deary, I., Weiss, A. & Batty, G. (2010). Intelligence and personality predictors of illness and death: How researchers in differential psychology and chronic disease epidemiology are collaborating to understand and address health inequalities. *Psychological Science in the Public Interest, 11(2)*, 52–79.

Eagly, A. (1987). *Sex Differences in Social Behavior: A Social Role Interpretation*. Hillsdale, NJ: Erlbaum.

Eagly, A. H. (2009). The his and hers of prosocial behavior: An examination of the social psychology of gender. *American Psychologist, 64*, 644–658.

Eagly, A. & Steffen, V. J. (1986). Gender and aggressive behavior: A meta-analytic review of the social psychological literature. *Psychological Bulletin, 100*, 309–330.

Eagly, A. & Wood, W. (1999). The origins of sex differences in human behavior: Evolved dispositions versus social roles. *American Psychologist, 54*, 408–423.

Eals, M. & Silverman, I. (1994). The hunter-gatherer theory of spatial sex differences: Proximate factors mediating the female advantage in recall of object arrays. *Ethology and Sociobiology, 15*, 95–105.

Easton, J., Schipper, I., & Schackleford, T. (2007). Morbid jealousy from an evolutionary psychological perspective. *Evolution and Human Behavior, 28*, 399–402.

Ekman, P. & Friesen, W. V. (1986). A new pan-cultural facial expression of emotion. *Motivation and Emotion, 10*, 159–167.

Ekman, P., Friesen, W. V., & Ellsworth, P. (1972). *Emotion in the Human Face: Guidelines for Research and a Review of Findings*. New York, NY: Permagon.

Ekman, P. (2001). *Telling lies: Clues to Deceit in the Marketplace, Politics, and Marriage* (3rd ed.). New York, NY: Norton.

Ekman, P. & Keltner, D. (1997). Universal facial expressions of emotion: An old controversy and new findings. In: U. Segerstrale, & P. Molnar (Eds.), *Nonverbal Communication: Where Nature Meets Culture*. Mahwah, NJ: Lawrence Erlbaum Associates.

Ekman, P. & O'Sulivan, M. (1991). Who can catch a liar? *American Psychologist, 46*, 913–920.

Ekman, P., O'Sullivan, M. & Frank, M. G. (1999). A few can catch a liar. *Psychological Science, 10*, 263–266.

Elfenbein, H. A. & Ambady, N. (2002). On the universality and cultural specificity of emotion. *Psychological Bulletin, 128(2)*, 203–235.

Elias, M. J. (1994). Capturing excellence in applied settings: A participant conceptualizer and praxis explicator role for community psychologists. *American Journal of Community Psychology, 22*, 293–318.

Ellsworth, P. C. & Carlsmith, J. M. (1973). Eye contact and gaze aversion in aggressive encounter. *Journal of Personality and Social Psychology, 33*, 117–122.

Emmons, R. A. & Paloutzian, R. F. (2003). The psychology of religion. *Annual Review of Psychology, 54*, 377–402.

Ferguson, C. (2013). Violent video games and the Supreme Court: Lessons for the scientific community in the wake of Brown v. Entertainment Merchants Association. *American Psychologist, 68*, 57–74.

Ferguson, C. (2012). Sandy Hook Shooting: Video Games Blamed, Again. ideas.time.com/2012/12/20/sandy-hook-shooting-video-games-blamed-again

Fischer, A. H. & Rodriguez Mosquera, P. (2001). What concerns men? Women or other men? *Psychology, Evolution & Gender, 3*, 5–25.

Fischer, P. & Greitemeyer, T. (2006). Music and aggression: The impact of sexual-aggressive song lyrics on aggression-related thoughts, emotions, and behavior toward the same and the opposite sex. *Personality and Social Psychology Bulletin, 32*, 1165–1176.

Fitch, W. T., Huber, L. & Bugnyar, T. (2010). Social cognition and the evolution of language: Constructing cognitive phylogenies. *Neuron, 65*, 795–814.

Flannery, D. J., Liau, A. K., Powell, K. E., Vesterdal, W., Vazsonyi, A. T., Guo, S., Atha, H., & Embry, D. (2003). Initial behavior outcomes for the PeaceBuilders Universal School-Based Violence Prevention Program. *Developmental Psychology, 39*, 292–308.

Folger, R. & Baron, R. A. (1996). Violence and hostility at work: A model of reactions to perceived injustice. In: G. R. Vanden Bos & E. Q. Bulato (Eds.), *Violence on the job: Identifying the risks and developing solutions* (pp. 51–85). Washington, DC: American Psychological Association.

Frank, M., Ekman, P., & Friesen, V. (1993). Behavioral markers and recognizability of the smile of enjoyment. *Journal of Personality and Social Behavior, (64)*, 83–93.

Frankl, V. (1959). *Man's Search for Meaning*. Boston: Beacon Press.

Freeman, M. (2012). Thinking and being otherwise: Aesthetics, ethics, erotics. *Journal of Theoretical and Philosophical Psychology, 32(4)*, 196–208.

Freud, S. (1930/1961). Civilization and its Discontents. New York, NY: W. W. Norton publishers.

Friederici, A. D., Mueller, J. L., & Oberecker, R. (2011). Precursors to natural grammar learning: Preliminatry evidence from 4-month olds. *PLoS One, 6*, e17920.

Friedman, H. & Booth-Kewley, S. (1987). Personality, type A behavior, and coronary heart disease: The role of emotional expression. *Journal of Personality and Social Psychology, 53(4)*, 783–792.

Fu, H., Watkins, D. & Hui, E. K. P. (2004). Personality correlates of the disposition towards interpersonal forgiveness: A Chinese perspective. *International Journal of Psychology, 39*, 305–316.

Gabriel, S. & Gardner, W. L. (1999). Are there "his" and "her" types of interdependence? The implications of gender differences in collective versus relational interdependence for affect, behavior, and cognition. *Journal of Personality and Social Psychology, 77,* 642–655.

Gall, T. L., Kristjansson, E., Charbonneau, C., & Florack, P. (2009). A longitudinal study on the role of spirituality in response to the diagnosis and treatment of breast cancer. *Journal of Behavioral Medicine, 32*, 132–186.

Gallup, G. G. & Frederick, D. A. (2010). The science of sex appeal: An evolutionary perspective. *Review of General Psychology, 14(3)*, 240–250.

Gangestad, S. W., Simpson, J. A., Cousins, A. J., Garver-Apgar, C. E., & Christensen, P. N. (2004). Women's preference for male behavioral displays changes across the menstrual cycle. *Psychological Science, 15*, 203–206

Geis, F. L. (1993). Self-fulfilling prophecies: A social psychological view of gender. In: A. E. Beall & R. J. Sternberg (Eds.), *The Psychology of Gender* (pp. 9–54). New York, NY: Guilford Press.

Gildersleeve, K. A., Haselton, M. G., Larson, C. M., & Pillsworth, E. G. (2012). Body odor attractiveness as a cue of impending ovulation in women: Evidence from a study using hormone-confirmed ovulation. *Hormones and Behavior, 61*, 157–166.

Goetz, J. L., Dacher, K., & Simon-thomas, E. (2010). Compassion: An evolutionary analysis and empirical review. *Psychological Bulletin, 136(3)*, 351–374.

Goodall, J. (1979). Life and death at Gombe. *National Geographic, 155*, 592–621.

Goodey, C. F. (1992). Mental disabilities and human values in Plato's Late Dialogues. *Archiv für Geschichte der Philosophie, 74(1)*, 26–42.

Gray, K., Ward, A. F., & Norton, M. I. (2014). Paying it forward: Generalized reciprocity and the limits of generosity. *Journal of Experimental Psychology, 143(1)*, 247–254.

Green, C. D. (2009). Darwinian theory, Functionalism, and the first American psychological revolution. *American Psychologist, 64*, 75–83.

Greenbaum, P. & Rosenfield, H. W. (1978). Patterns of avoidance in responses to interpersonal staring and proximity: Effects of bystanders on drivers at a traffic intersection. *Journal of Personality and Social Psychology, 36*, 575–587.

Greenwood, D. N. (2013). Fame, Facebook, and twitter: How attitudes about fame predict frequency and nature of social media use. *Psychology of Popular Media Culture, 2(4)*, 222–236.

Greitemeyer, T., Osswald, S. & Brauer, M. (2010). Playing prosocial video games increases empathy and decreases schadenfreude. *Emotion, (10)*, 796–802.

Gruber, J., Quoidbach, J., Kogan, A. & Mauss, I. (2013). Happiness is best kept stable: Positive emotion variability is associated with poorer psychological health. *Emotion, 13*, 1–6.

Hall, J. A., Murphy, N. A., & Schmid Mast, M. (2006). Recall of nonverbal cues: Exploring a new definition of interpersonal sensitivity. *Journal of Nonverbal Behavior, 30*, 141–155.

Hamilton, W. D. (1964). The genetic evolution of social behavior. *Journal of Theoretical Biology, (7)*, 1–16.

Hamlin, J. K., Wynn, K., & Bloom, P. (2007). Social evaluation by preverbal infants. *Nature, 450*, 557–559.

Haselton, M. G. & Gildersleeve, K. (2011). Can men detect ovulation? *Current Directions in Psychological Science, 20*, 87–92.

Hastings, E., Karas, T., Winsler, A., Way, E., Madigan, A., & Tyler, S. (2009). Young children's video/computer game use: Relations with school performance and behavior. *Issues in Mental Health Nursing, 30*, 638–649.

Hayes, S. C., Rincover, A. & Volosin, D. (1980). Variables influencing the acquisition and maintenance of aggressive behavior: Modeling versus sensory reinforcement. *Journal of Abnormal Psychology, 89*, 254–262.

Henderson, J. A. & Anglin, J. M. (2003). Facial attractiveness predicts longevity. *Evolution and Human Behavior, 24*, 351–356.

Heyes, C. (2012). What's social about social learning? *Journal of Comparative Psychology, 126*, 193–202.

Hill, K. & Hurtado, A. M. (1996). *Ache Life History*. New York, NY: Aldine DeGruyer.

Hilton, N. Z., Harris, G. T., & Rice, M. E. (2000). The functions of aggression by male teenagers. *Journal of Personality and Social Psychology, 79*, 988–994.

Hobbes, T. (1651). *Leviathon*. New York, NY: Oxford University Press.

Hoffman, A. J., Wallach, J., Espinoza Parker, N. & Sanchez, E. (2009). *Unity through community service activities: Strategies to build ethnic and cultural divides*. McFarland Publishers.

Hoffman, A. J., Wallach, J. N., & Sanchez, E. (2010). Community service work, civic engagement and "giving back" to society: Key factors in improving interethnic relationships and achieving "connectedness" in ethnically diverse communities. *Australian Social Work, 63(4)*, 418–430.

Horne, A. M., Stoddard, J. L., & Bell, C. D. (2007). Group approaches to reducing aggression and bullying in school. *Group Dynamics, Theory, Research and Practice, 11*, 262–271.

Hovland, C. J. & Sears, R. R. (1940). Minor studies in aggression: VI. Correlation of lynchings with economic indices. *Journal of Psychology, 9*, 301–310.

Hubard, J. (2001). Emotion expression processes in children's peer interaction: The role of peer rejection, aggression and gender. *Child Development, 72*, 1426–1438.

Huesmann, L. R. (2010). Nailing the coffin shut on doubts that violent video games stimulate aggression: Commentary on Anderson, et al. *Psychological Bulletin, 136*, 179–181.

Huesmann, L. R. (2007). The impact of electronic media violence: Scientific theory and research. *Journal of Adolescent Health, 41(6 Suppl 1)*, S6–S13.

Izard, C. (1991). *The psychology of Emotions*. New York, NY: Plenum.

James, W. (1950). *Principles of Psychology, Vol. 1*. New York, NY: Dover (original work published 1890).

Janson, C. H. & van Shaik, C. P. (2000). The behavioral ecology of infanticide by males. In: C. P. Shaik, & C. H. Hanson (Eds.), *Infanticide in males and its implications* (pp. 469–494). Cambridge, MA: Cambridge University Press.

Jason, L. (2006). Benefits and challenges of generating community participation. *Professional Psychology: Research and Practice, 37*, 132–139.

Johnson, C. (1990). *The Western Political Quarterly, 43(4)*, 719–740.

Johnson, D. W. & Johnson, R. T. (1975). *Learning Together and Alone: Cooperation, Competition, and Individualization*. Englewood Cliffs, NJ: Prentice-Hall.

Johnson, W., Johnson, R. T., Johnson, J. & Anderson, D. (1976). Effects of cooperative versus individualized instruction on student prosocial behavior, attitudes toward learning, and achievement. *Journal of Educational Psychology, 68(4)*, 446–452.

Junco, R. (2012). Too much face and not enough books: The relationship between multiple indices of Facebook use and academic performance. *Computers in Human Behavior, 28*, 187–198.

Kaag, J. & Bhatia, S. (2014). Fools for tools: Why engineers need to become philosophers. *The Chronicle of Higher Education, November 28* (Section B), B13–B15.

Kagan, S. (1992). *Cooperative Learning*. San Juan Capistrano, CA: Resources for Teachers, Inc.

Karremans, J. C. & Van Lange, P. A. M. (2004). Back to caring after being hurt: The role of forgiveness. *European Journal of Social Psychology, 34*, 207–227.

Kim, S., Kamphaus, R., Orpinas, P., & Kelden, S. H. (2011). A multiple risk factors model of the development of aggression among early adolescents from urban disadvantaged neighborhoods. *School Psychology Quarterly, 26*, 215–230.

Kimble, G. A. (1994). A new formula for Behaviorism. *Psychological Review, 101*, 254–258.

Kinder, D. R. & Sears, D. O. (1981). Prejudice and politics: Symbolic racism versus racial threats to the good life. *Journal of Personality and Social Psychology, 40*, 414–431.

Kirsh, S. J. (2012). *Children, Adolescents, and Media Violence: A Critical Look at the Research* (2nd ed.). Thousand Oaks, CA: Sage.

Krieger, L. (2015). Bigger is better, evolutionarily. *Star Tribune*, Science & Health, Sunday March 1.

Kruger, D. J., Fisher, M. L., & Wright, P. (2013). A framework for integrating evolutionary and Feminist perspectives in psychological research. *Journal of Social, Evolutionary and Cultural Psychology, 7(4)*, 299–303.

Krumhuber, E., Manstead, A. S., Cosker, D., Marshall, D., Kappas, A., & Rosin, P. (2007). Facial dynamics as indicators of trustworthiness and cooperative behavior. *Emotion, 7(4)*, 730–735.

Kuhl, P. K. & Damasio, A. (2011). Language. In: E. R. Kandel & others (Eds), *Principles of Neural Science* (5th ed.). New York, NY: McGraw-Hill.

Landy, F. J. (1997). Early influences on the development of industrial and organizational psychology. *Journal of Applied Psychology, 82(4)*, 467–477.

Leal, S. & Vrij, A. (2008). Blinking during and after lying. *Journal of Nonverbal Behavior, 32*, 187–194.

Lester, D. (1991). *Questions and Answers about Murder*. Philadelphia, PA: Charles Press.

Liddle, J. R., Shackelford, T. K., & Weeks-Shackelford, V. A. (2012). Why can't we all just get along? Evolutionary perspectives on violence, homicide, and war. *Review of General Psychology, 16*, 24–36.

Lock, A., & Zukow-Goldring, P. (2010). Preverbal communication. In: J. G. Bremner & T. D. Wachs (Eds.), *Wiley-Blackwell Handbook of Infant Development* (2nd ed.). New York, NY: Wiley.

Locke, J. (1963). *An Essay Concerning Human Understanding*. Germany: Scientia Verlag Aalen. (Original work published in 1690).

Loftus, E. F. (2013). Eyewitness testimony in the Lockerbie bombing case. *Memory, 21*, 584–590.

Maccoby, E. E. (1990). Gender and relatinships: A developmental account. *American Psychologist, 45*, 513–520.

Maccoby, E. E. & Jacklin, C. N. (1987). Gender segregation in childhood. In: H. W. Reese (Ed.), *Advances in Child Development and Behavior* (Vol. 20, pp. 239–287). New York, NY: Cambridge University Press.

Machery, E. (2011). A better philosophy for a better psychology: Comment on Slaney and Racine (2011). *Journal of Theoretical and Philosophical Psychology, 31(2)*, 90–95.

Markus, H. R. & Kitayama, S. (1991). Culture and the self: Implications for cognition, emotion, and motivation. *Psychological Review, 98(2)*, 224–253.

Marsh, A. A., Kozak, M. N., & Ambady, N. (2007). Accurate identification of fear facial expressions predicts prosocial behavior. *Emotion, 7*, 239–251.

McKnight, D. H., Cummings, L. L., & Chervang, N. L. (1998). Initial trust formation in new organizational relationships. *The Academy of Management Review, 23*, 473–490.

McCullough, M. E. (2008). *Beyond Revenge: The Evolution of the Forgiveness Instinct*. John Wiley & Sons.

McCullough, M. E., Berry, J. W., Root Luna, L., Tabak, B. A. & Bono, G. (2010). On the form and function of forgiving: Modelilng the time-forgiveness relationship and testing the valuable relationships hypothesis. *Emotion, 10*, 358–376.

Miles, S. J. (2001). Charles Darwin and Asa Gray discuss teleology and design. *Perspectives on Science and Christian Faith, 53(3)*, 196–201.

Miller, S. L., Maner, J. K. & Becker, D. V. (2010). Self-protective biases in group categorization: Threat cues shape the psychological boundary between "us" and "them." *Journal of Personality and Social Psychology, 99*, 62–77.

Mintz, A. (1946). A re-examination of correlations between lynchings and economic indices. *Journal of Abnormal Psychology, 41*, 154–160.

Mouch, C. A. & Sonnega, A. J. (2012). Spirituality and recovery from cardiac surgery: A review. *Journal of Religious Health, 51*, 1042–1060.

Mullet, E., Girard, M. & Bakhshi, P. (2004). Conceptualizations of forgiveness. *European Psychologist, 9*, 78–86.

Nails, D. (2014). Socrates, The Stanford Encyclopedia of Philosophy (Spring 2014). URL = <http://plato.stanford.edu/archives/spr2014/entries/socrates/>.

Nowak, M. A. (2012). Why we help: The evolution of cooperation. http://www.scientificamerican.com/article.cfm?id=why-we-help-evolution-cooperation

O'Connor, S. & Jose, P. E. (2012). A propensity score matching study of participation in community activities: A path to positive outcomes for youth in New Zealand? *Developmental Psychology, 48*, 1563–1569.

Ohbuchi, K. & Kambara, T. (1985). Attacker's intent and awareness of outcome, impression management, and retaliation. *Journal of Experimental Social Psychology, 21*, 321–330.

Orians, G. H. & Heerwagen, J. H. (1992). Evolved responses to landscapes. In: J. Barkow, L. Cosmides, & J. Tooby (Eds.), *The Adapted Mind* (pp. 55–579). New York, NY: Oxford University Press.

Paik, H. & Comstock, G. (1994). The effects of television violence on antisocial behavior: A meta-analysis. *Communication Research, 21*, 516–546.

Parens, H. (2014). *War is Not Inevitable*. Lanham, Maryland: Lexington Books.

Pargament, K. I. (1997). *The Psychology of Religion and Coping*. New York, NY: Guilford press.

Park, J. H., Shaller, M. & Van Vugt, M. (2008). Psychology of human kin recognition: Heuristic cues, erroneous inferences, and their implications. *Review of General Psychology, 12*, 215–235.

Parkinson, B. (2005). Do facial movements express emotions or communicate motives? *Personality and Social Psychology Review, 9*, 278–311.

Pennypacker, H. S. (1992). Is behavior analysis undergoing selection by consequences? *American Psychologist, 47*, 1491–1498.

Peterson, C. & Seligman, M. E. P. (2004). *Character Strengths and Virtues*. Oxford: Oxford University Press.

Pettigrew, T. F. (1998). Intergroup contact theory. *Annual Review of Psychology, 49*, 65–85.

Pettigrew, T. F. & Tropp, L. R. (2006). A meta-analytic test of intergroup contact theory. *Journal of Personality and Social Psychology, 90*, 751–783.

Piaget, J. (1932). *The Moral Judgment of the Child*. New York, NY: Harcourt, Brace & World.

Pinker, S. (1994). *The Language Instinct*. New York, NY: Harper Collins.

Pinker, S. & Bloom, P. (1990). Natural language and natural selection. *Behavioral and Brain Sciences, 13*, 707–784.

Putnam, R. (2000). *Bowling alone: The Collapse and Revival of American Community*. New York, NY: Simon & Shuster.

Pressman, S. D., Gallagher, M. W., Lopez, S. J. & Campos, B. (2014). Incorporating culture into the study of affect and health. *Psychological Science, 25(12)*, 2281–2283.

Rapoport, A. & Chammah, A. M. (1965). Sex differences in factors contributing to the level of cooperation in the prisoner's dilemma game. *Journal of Personality and Social Psychology, 2*, 831–838.

Reed, A. & Aquino, K. F. (2003). Moral identity and the expanding circle of moral regard toward out-groups. *Journal of Personality and Social Psychology, 84*, 1270–1286.

Reiber, C. & Garcia, J. R. (2010). Hooking Up: Gender differences, evolution, and pluralistic ignorance. *Evolutionary Psychology, 8(3)*, 390–404.

Rensch, B. (1948). Historical changes correlated with evolutionary changes of body size. *Evolution, (3)*, 218–230.

Richards, R. (2008). *This Tragic Sense of Life: Ernst Haeckel and the Struggle over Evolutionary Thought*. Chicago: University of Chicago Press.

Rusiniak, K., Garcia, J. & Hankins, W. (1976). Flavor illness aversions: Potentiation of odor by taste in rats. *Journal of Comparative Physiological Psychology, 90*, 460–467.

Rymer, R. (1993). *Genie: A Scientific Tragedy*. New York, NY: Harper Perennial.

Sacco, D. F. & Hugenberg, K. (2009). The look of fear and anger: Facial maturity modulates recognition of fearful and angry expressions. *Emotion, 9*, 39–49.

Sarason, S. B. (1974). *The Psychological Sense of Community: Prospects for Community Psychology*. San Francisco: Jossey-Bass.

Schat, A. C. H. & Kelloway, K. E. (2003). Reducing the adverse consequences of workplace aggression and violence: The buffering effects of organizational support. *Journal of Occupational Health Psychology, 8*, 110–122.

Sears, D. O. & Kinder, D. R. (1985). Whites' opposition to busing: On conceptualizing and operationalizing group conflict. *Journal of Personality and Social Psychology, 48*, 1141–1147.

Schmid Mast, M. & Hall, U. A. (2006). Women's advantage at remembering others' appearance: A systematic look at the why and when of a gender difference. *Personality and Social Psychology Bulletin, 32*, 353–364.

Schmitt, D. P. & Buss, D. M. (1996). Strategic self-promotion and competitor derogation: Sex and context effects on perceived effectiveness of mate attraction tactics. *Journal of Personality and Social Psychology, 70*, 1185–1204.

Schofield, M. (2006). *Plato: Political Philosophy*. Oxford University Press

Shaver, P. R., Murdaya, U., & Fraley, R. C. (2001). Is love a "basic" emotion? *Personal Relationships, 3*, 81–96.

Shields, S. A. & Bhatia, S. (2009). Darwin on race, gender and culture. *American Psychologist, 64*, 111–119.

Simon, B. (1972). Models of mind and mental illness in ancient Greece: II. The Platonic model. *Journal of the History of the Behavioural Sciences, 8*, 389–404.

Simon, B. (1973). Models of mind and mental illness in ancient Greece: II. The Platonic model: Section 2. *Journal of the History of the Behavioral Sciences, 8*, 389–404.

Singh, D. (1993). Adaptive significance of female attractiveness: Role of waist-hip-ratio. *Journal of Personality and Social Psychology, 65*, 293–307.

Skinner, B. F. (1938). *The Behavior of Organisms*. New York, NY: Appleton-Century-Crofts.

Skinner, B. F. (1948). *Walden Two*. New York, NY: Macmillan.

Skinner, B. F. (1953). *Science and Human Behavior*. New York, NY: Free Press.

Skinner, B. F. (1957). *Verbal Behavior*. Acton, MA: Copley Publishing Group.

Skinner, B. F. (1981). Selection by consequences. *Science, 213*, 501–504.

Skinner, B. F. (1984). The shame of American education. *American Psychologist, 39*, 947–954.

Skon, L., Johnson, D. W., & Johnson, R. T. (1981). Cooperative peer interaction versus individual competition and individualistic efforts: Effects on the acquisition of cognitive reasoning strategies. *Journal of Educational Psychology, 73*, 83–92.

Slavin, R. E. (1990). *Cooperative Learning: Theory, Research and Practice*. Englewood Cliffs, NJ: Prentice Hall.

Smith, E. A. (2010). Communication and collective action: Language and the evolution of human cooperation. *Evolution and Human Behavior, 10*, 231–245.

Stark, S. A. (2014). Implicit virtue. *Journal of Theoretical and Philosophical Psychology, 34*, 146–158.

Sterelny, K., Joyce, K., Calcott, B. & Fraser, B. (2013). *Cooperation and Its Evolution*. Cambridge, MA: The MIT Press.

Suddendorf, T. (2013). Where are Humans' Close Cousins? *The Chronicle of Higher Education, 60(8)*, B4–B5.

Sullaway, M. (2004). Psychological perspectives on hate crime laws. *Psychology, Public Policy and Law, 10*, 250–292.

Sundie, J. M., Griskevicius, V., Vohs, K. D., Kendrick, D. T., Tybur, J. M., & Beal, D. J. (2011). Peacocks, porsches, and thorstein Veblen: Conspicuous consumption as a sexual signaling system. *Journal of Personality and Social Psychology, 100*, 664–680.

Symons, D. (1995). Beauty is in the adaptiveness of the beholder. In: P. R. Abramson & S. D. Pinkerson (Eds.), *Sexual Nature, Sexual Culture* (pp. 80–118). Chicago: University of Chicago Press.

Tavris, C. (1989). *Anger: The Misunderstood Emotion* (2nd ed.). New York, NY: Simon & Schuster.

Thorpe, S. K., Holder, R. L., & Crompton, R. H. (2007). Origin of human bipedalism as an adaptation for locomotion on flexible branches. *Science, 316*, 1328–1331.

Tidball, K. G., Krasny, M. E., Svendsen, E., Campbell, L., & Helphand, K. (2010). Stewardship, learning, and memory in disaster resilience. *Environmental Education Research, 16*, 591–609.

Tooby, J. & DeVore, I. (1987). The reconstruction of hominid behavioral evolution through strategic modeling. In: W. G. Kinzey (Ed.), *The Evolution of Human Behavior* (pp. 183–237). New York, NY: State University of New York Press.

Toussaint, L. L., Owen, A. D., & Cheadle, A. (2012). Forgive to live: Forgiveness, health and longevity. *Journal of Behavioral Medicine, 35*, 375–386.

Trivers, R. (1971). The evolution of reciprocal altruism. *Quarterly Review of Biology, 46*, 35–57.

Trivers, R. (1972). Parental investment and sexual selection. In: B. Campbell (Ed.), *Sexual Selection and the Descent of Man: 1871–1971* (pp. 136–179). Chicago: Aldine Press.

Tyler, J. M., Feldman, R. S., & Reichert, A. (2006). The price of deceptive behavior: Disliking and lying to people who lie to us. *Journal of Experimental Social Psychology, 42*, 69–77.

Unger, R. K. (2014). It is hardly news that women are oppressed: Sexism, activism and *Charlie. Journal of Cognitive Neuroscience, 26(6)*, 1324–1326.

Vanello, J. A. & Cohen, D. (2003). Male honor and female infidelity: Implicit cultural scripts that perpetuate domestic violence. *Journal of Personality and Social Psychology, 84*, 997–1010.

Van Hoof, M. H., Voorhoost, F. J., Kaptein, M. B., Hirasng, R. A., Koppenaal, C., & Schoemaker, J. (2000). Insulin, androgen, and gonadotropin concentration, body mass index, and waist-to-hip ratio in the first years after menarche in girls with regular menstrual cycle, irregular menstrual cycles, or oligomenorrhea. *Journal of Clinical Endocrinology and Metabolism, 85*, 1394–1400.

Van Vugt, M. & Schaller, M. (2008). Evolutionary approaches to group dynamics: An introduction. *Group Dynamics: Theory, Research and Practice, 12*, 1–6.

Vick, S. J., Waller, B. M., Parr, L. A., Smith-Pasqualini, M. C., & Bard, K. A. (2006). A Cross-species comparison of facial morphology and movement in humans and chimpanzees using the facial action coding system. *Journal of Nonverbal Behavior, 31,* 1–16.

Voracek, M., & Fisher, M. L. (2006). Success is all in the measures: Androgenousness, curvaceousness, and starring frequencies in adult media actresses. *Archives of Sexual Behavior, 35*, 297–304.

Vygotsky, L. S. (1978). *Mind in Society: The Development of Higher Psychological Processes*. Cambridge, MA: Harvard University Press.

Walsh, T. G., Teo, T., & Baydala, A. (2014). *A Critical History and Philosophy of Psychology*. New York, NY: Cambridge University Press.

Warneken, F., & Tomasello, M. (2009). Varieties of altruism in children and chimpanzees. *Trends in Cognitive Sciences, 13*, 397–402.

Watson, J. B. (1924). *Behaviorism*. New York, NY: People's Institute.

Webster, G. D. (2003). Prosocial behavior in families: Moderators of resource sharing. *Journal of Experimental Social Psychology, 39*, 644–652.

Webster, G. D. (2008). The kinship, acceptance, and rejection model of altruism and aggression (KARMAA): Implications for interpersonal and intergroup aggression. *Group Dynamics, Theory, Research and Practice, 12,* 27–38.

Weiner, J. (1994). *The Beak of the Finch*. New York, NY: Vintage Books.

Wilkinson, G. W. (1984). Reciprocal food sharing in the vampire bat. *Nature, 308*, 181–184.

Wilson, M. & Daly, M. (1985). Competitiveness, risktaking, and violence: The young male syndrome. *Ethology and Sociobiology, 6*, 59–73.

Wilson, M. & Daly, M. (2004). Do pretty women inspire to discount the future? *Proceedings of the Royal Society B: Biological Sciences, 271(Suppl. 4)*, S177–S179.

Wolff, T. (2014). Community psychology practice: Expanding the impact of psychology's work. *American Psychologist, 69(8)*, 803–813

Wood, A. & Eagly, A. H. (2010). Gender. In: S. T. Fiske, D. T. Gilbert, & G. Lindzey, (Eds.), *Handbook of Social Psychology* (Vol. 1, 5th ed., pp. 629–667). New York, NY: Wiley.

Wood, D., Bruner, J. S., & ross, G. (1976). The role of tutoring in problem solving. *Journal of Child Psychology and Psychiatry, 17*, 89–100.

Wood, W., Wong, F. Y., & Cachere, J. G. (1991). Effects of media violence on viewers' aggression in unconstrained social interaction. *Psychological Bulletin, 109*, 371–383.

Woodward, A., Markman, E., & Fitzsimmons, C. (1994). Rapid words learning in 13- and 18-month-olds. *Developmental Psychology, 30*, 553–556.

Wundt, W. (1904). *Principles of Physiological Psychology* (5th ed.) (E. B. Titchener, Trans.). New York, NY: Macmillan. (Original work published 1873–1874).

Yamagishi, T. (2011). *Trust: The Evolutionary Game of Mind and Society*. New York, NY: Springer.

Yamamoto, S. & Takimoto, A. (2012). Empathy and fairness: Psychological mechanisms for eliciting and maintaining prosociality and cooperation in primates. *Social Justice Research, 25*, 233–255.

Yerkes, R. M. (1925). *Almost Human*. London, England: Random House UK.

Young-Browne, G., Rosenfeld, H. M., & Horowitz, F. D. (1977). Infant discrimination of facial expressions. *Child Development, 48*, 555–562.

Zaki, J. & Mitchell, J. P. (2013). Intuitive prosociality. *Current Directions in Psychological Science, 22*, 466–470

Zivin, G. (1982). Watching the sands shift: Conceptualizing the development of nonverbal mastery. In: R. S. Feldman (Ed.), *Development of Nonverbal Behavior in Children* (pp. 73–98). New York, NY: Springer-Verlag.

Index

About the Author

August John Hoffman is currently a Professor of Psychology at Metropolitan State University. He earned his BA from UC Santa Barbara, MA from Radford University in Clinical Psychology (with an emphasis in Sport Psychology), and PhD from UCLA in counseling psychology. As a former professor of psychology at Compton College, CSU Northridge, and Pepperdine University, he has assisted students from various educational backgrounds in accomplishing their goals. He began and developed a highly successful gardening program at Compton College in an effort to help students improve their campus and community. More recently Dr. Hoffman has explored the psychological and community benefits of community gardening and has helped organize community gardens throughout the United States (Detroit, MI, Newtown Victory Garden near the Sandy Hook Elementary School, and Red Lake, MN Tribal Nation) and internationally (rural villages in Guatemala). Dr. Hoffman's current research projects at Metropolitan State University include urban forestry and the development of a community fruit tree orchard and community garden at Inver Hills Community College with Dr. Barbara Curchack. Current research interests also include community service work and student mentoring as effective methods to reduce ethnic conflict and improve social capital among student and community members. For the last five years Dr. Hoffman has conducted research combining outdoor gardening work with Metropolitan State University students with community members. Students generally appreciate the ability to engage in a variety of community service projects and see how psychological theory applies itself within the community with community members. Additionally, Dr. Hoffman has taught several psychology courses including Community Psychology, Evolutionary Psychology and Sport Psychology, which include an applied approach to creating healthy lifestyles for his students. He has published several books and academic research articles, including the texts, *Unity Through Community Service Activities*; *Understanding Sport Psychology and Human Behavior*; and, *Stop Procrastinating Now!*

He enjoys gardening at his home in Hudson, Wisconsin, during his time off with his family—Nancy his wife, and two children A. J. and Sara.

CPSIA information can be obtained
at www.ICGtesting.com
Printed in the USA
LVHW031801291220
675341LV00002B/388

9 781498 518178